Dig In!
Hands-On Soil Investigations

Dig In!

Hands-On Soil Investigations

Featuring *sci*LINKS®—a new way of connecting text and the Internet.
Up-to-the-minute online content, classroom ideas, and other materials are
just a click away. Go to page xxii to learn more about this educational
resource.

NATIONAL SCIENCE TEACHERS ASSOCIATION

Arlington, Virginia

Published with support from the Natural Resources Conservation Service,
U.S. Department of Agriculture

Shirley Watt Ireton, Director
Beth Daniels, Managing Editor
Judy Cusick, Associate Editor
Jessica Green, Assistant Editor
Linda Olliver, Cover Design

Art and Design
Linda Olliver, Director
NSTA Web
Tim Weber, Webmaster
Periodicals Publishing
Shelley Carey, Director
Printing and Production
Catherine Lorrain-Hale, Director
Publications Operations
Erin Miller, Manager
*sci*LINKS
Tyson Brown, Manager

National Science Teachers Association
Gerald F. Wheeler, Executive Director
David Beacom, Publisher

NSTA Press, NSTA Journals,
and the NSTA Website deliver
high-quality resources for
science educators.

Dig In! Hands-On Soil Investigations
NSTA Stock Number: PB159X
ISBN 0-87355-189-3
Library of Congress Card Number: 00-112121
Printed in the USA by FRY COMMUNICATIONS, INC.
Printed on recycled paper

Contents

Section I
What Is Soil?

Section II
Who Uses Soil?

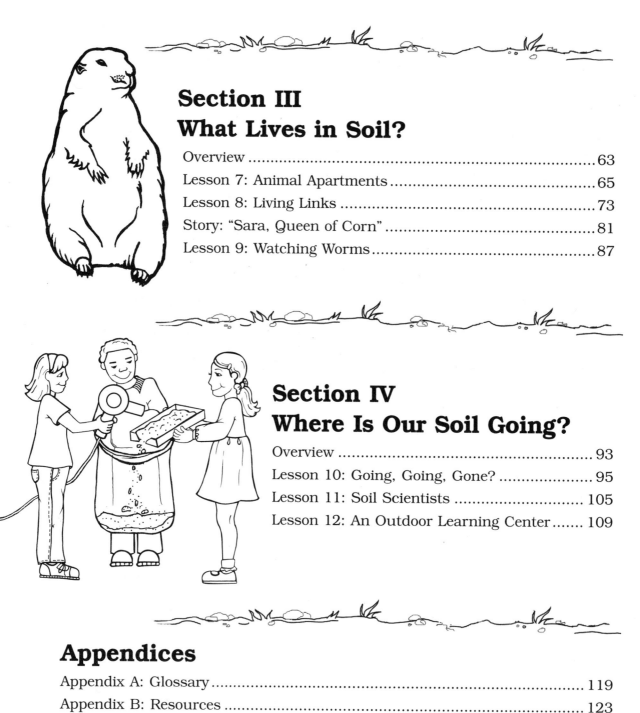

Section III
What Lives in Soil?

Section IV
Where Is Our Soil Going?

Appendices

Preface

Nature offers many moods: the serenity of running water, the seasonal color change of leaves, the violence of a howling blizzard, the movement of animals in flight, the pastoral beauty of a landscape, the kaleidoscope of colors in a desert sunset. But as we enjoy the natural world, we don't often think of one very important part: the soil. Soil is the substance in which most of our food is grown or raised. Soil is also the mud that squishes under our shoes after a rainstorm, and the grit that grazes our faces as the wind whips across a dry landscape. Soil provides space for our buildings and communities, but also is the substance that muddies our rivers when we don't properly care for the land.

Soil results from a complex series of geological, human, and biological forces. It is a tangible and traceable record of these forces. This record is illustrated by the color, feel, odor, and even the taste differences of the world's thousands of soils.

The Natural Resources Conservation Service (NRCS) has collaborated with the National Science Teachers Association (NSTA) on *Dig In! Hands-On Soil Investigations* for elementary science teachers and supervisors. The activities in *Dig In!*, designed for students in kindergarten through fourth grade, introduce soil's mysteries in an enjoyable and educational way.

As you use these activities, you and your students will gain a greater appreciation for the value of soil. Such an understanding is critical if today's students are to become informed decision-makers and conservers of our natural resources.

Acknowledgments

Dig-In! Hands-On Soil Investigations is a co-publication of the Natural Resources Conservation Service (an agency of the U.S. Department of Agriculture) and the National Science Teachers Association.

This book was conceptualized by the former NRCS Educational Relations Staff and by Agri-Education, Stratford, Iowa, under contract with NRCS. Paul DuMont and Theodore Kupelian (Educational Relations Staff) and Thomas Levermann (NRCS) incorporated new ideas and product direction. Other former agency staff contributing to the initial development of *Dig-In!* include Hubert Kelly and Duane Bosworth. Lesson 9: Watching Worms was adapted from a *Science and Children* article by Lori Gibb (Noah Wallace School, Farmington, Connecticut). Sandra Laskey wrote and illustrated two special stories for teachers to read to students.

NRCS soil scientists, including Dr. Richard Arnold, Dr. Hari Eswaran, Dr. Sheryl Kunickis, and Soil Survey Division Director Horace Smith, reviewed various drafts of *Dig-In!* NRCS Earth Team volunteers contributed to the further development and refinement of *Dig-In!* Greg Donaldson developed and refined the initial publication design, and Barbara Levermann provided extensive proofreading and editorial review.

Dig-In! was thoroughly reviewed and tested by educators. In addition to the numerous educators who advised NRCS during the book's development, the book was reviewed by Rhonda Bajalia (Crown Point Elementary School, Jacksonville, Florida), Betsy Benz (Wickliffe Elementary School, Wickliffe, Ohio), Dr. E. Barbara Klemm (Associate Professor of Education, University of Hawaii at Manoa), and David Brown (St. Peter School, Quincy, Illinois). Special thanks also go to Melody Orban (Elementary Science Resource Teacher, Kenosha Unified School District, Kenosha, Wisconsin) and the following teachers for testing these activities with their students: Kimberly George, Mary Pilot, and Steve Plato (Bain Elementary School, Kenosha, Wisconsin); Judith Herr (Grewenow Elementary School, Kenosha, Wisconsin); Corinne Nelson and Debbie Schuebel (Harvey Elementary School, Kenosha, Wisconsin); Kathy Leffler (Jefferson Elementary School, Kenosha, Wisconsin); and Gigi Bohm and Sharon Tilton (Union Grove Elementary School, Union Grove, Wisconsin).

The NRCS project manager for *Dig In! Hands-On Soil Investigations* was Thomas Levermann, Head of Education and Publications, Conservation Communications Staff. At NSTA, the project editor was Jessica Green. Also at NSTA, Linda Olliver designed the book and the cover, Tracey Shipley and Joanne Cunha created line art, Nguyet Tran did book layout, and Catherine Lorrain-Hale coordinated production and printing of the book.

How to Use This Book

These activities are designed to heighten student awareness of the value of soil. The focus of information and activities should not be on getting the right answers, but on asking the right questions. All reasonable ideas given by students are acceptable.

Concepts and Vocabulary

Significant words are italicized in the Teacher Background of each lesson and are also defined in the glossary (Appendix A). These key words allow classroom teachers to match the skill and instructional areas required by school boards with activities contained in each lesson. You will also find tables on the following pages correlating each lesson with other disciplines, the *National Science Education Standards,* and the *Benchmarks for Science Literacy.*

Adapt the teaching methods in each lesson to your classroom's individual needs. You might use vocabulary lists, word charts, and concept maps to help students relate ideas and understand key terms.

Planning

Each lesson has a special emphasis and builds upon previous lessons, although each may be used separately. The lessons include background information and guidelines for conducting the activities. Each lesson is made up of five short activities that correspond to the stages in a student's learning cycle. The *Dig In!* learning cycle is adapted from the 5 E instructional model (Trowbridge and Bybee 1995).

1 *Perception:* students discuss ideas
2 *Exploration:* students engage in hands-on investigations of concepts
3 *Application:* students communicate ideas and apply ideas to a new situation
4 *Evaluation:* students' knowledge is assessed
5 *Extensions* (optional): students expand their understanding of concepts

Each Learning Cycle activity requires approximately 30 minutes. The directions for planning and conducting activities include estimated times, but the actual time will vary depending on your pupils' age and abilities and on material availability. Make sure students have plenty of time to explore and experiment, especially during the Exploration and Application stages.

Students can work individually or in groups. Dividing your class into groups will reduce the amount of materials and preparation time needed, and may help students learn the concepts. When conducting activities in groups, remind the class that everybody has strengths and weaknesses and that each group member should participate, cooperate, and contribute to the success of the group.

Think about ways to make the activities and learning fully accessible to all students, including those with special needs.

Materials

The activities call for inexpensive, low-impact items such as plastic jars, lids, egg cartons, and rocks. Before introducing each lesson, ask students to bring materials from home and send a note to parents explaining why these items are needed. Use recycled products and reuse items as much as possible. The materials list in each lesson gives the items necessary for the first four activities of the Learning Cycle, but not for the optional Extensions at the end of each lesson.

Soil is required for Lessons 1, 2, 3, 6, 9, and 10. In some areas of the United States, soil use is restricted because of concern about transporting agricultural pests and invasive species into vulnerable ecological areas. Check local regulations concerning the use and transportation of soil before you conduct these lessons. Ask the school custodian, a greenhouse or nursery, construction business, or a local NRCS or USDA Office for donations of or suggestions on how to obtain samples of silty, sandy, and clayey soil. In Lessons 1 and 2, your class will sample soil from the school yard. Before you start these lessons, secure permission from school administrators and custodial staff to take soil from the school grounds.

Do not substitute potting soil—because it has been sterilized, potting soil does not contain the items found in natural soil.

Only once, in Lesson 1, is it appropriate to use the word "dirt." After that, the proper term is "soil."

Stories

Lessons 6 and 8 include stories to read to students as an optional Extension activity. The stories are preceded by suggestions for activities that will reinforce the concepts in the stories and the lessons.

National Science Teachers Association

Maps for Learning

What should students learn? In what order? And how does each strand of knowledge connect to other vital threads? These are the tough questions every teacher faces, and the illustrations on the following pages are designed to help you answer them.

The following illustrations introduce a way of considering and organizing science content standards. The maps use the learning goals of the American Association for the Advancement of Science (AAAS)'s *Science for All Americans* and *Benchmarks for Science Literacy*. The maps are excerpted from *Atlas of Science Literacy,* copublished by AAAS and NSTA in a two-volume work (Volume I, AAAS 2001). The complete *Atlas* will contain nearly one hundred similar maps on the major elementary and secondary basic science topics: solar system, cells and organs, laws of motion, chemical reactions, evolution, and more.

How to Use the *Atlas* Maps

Atlas of Science Literacy shows that understanding one goal contributes to the understanding of another. Each *Atlas* map is designed to help clarify the context of a science benchmark or standard: where it comes from, where it leads, and how it relates to other standards. With the maps as guides, you can make sure your students have experience with the prerequisite learning, and you can draw students' attention to related content—getting their framework for learning ready!

The *Atlas* maps included in this book list the ideas relevant to students' understanding of three main topics: "Processes that Shape the Earth," "Flow of Matter in Ecosystems," and "Agricultural Technology" (see Figures i.1–i.3). These maps trace the ideal development of knowledge from kindergarten to eighth grade. Horizontal lines represent the grade-level appropriateness. Goals that deal with the same idea are organized into columns (called "strands") with more sophisticated goals above simpler ones, connected by arrows. The concept squares in the map that relate directly to *Dig In!* learning goals are shaded in gray. These are the concepts that you want your students to fully grasp. They underlie the Student Objectives presented at the beginning of each lesson.

The *Atlas* maps can help you connect your instruction to your state science standards. As of this writing, 49 of the 50 states in the US have developed their own standards, most modeled directly on the *Standards* or AAAS's

Benchmarks. The correlation between the *Standards* and *Benchmarks* in science content is nearly one hundred percent. So there is a unity of purpose and direction, if not quite a common language. Fortunately, the National Science Foundation, the Council of Chief State School Officers, and other groups have developed websites to guide educators in correlating these national standards with their state goals (for example, the ExplorAsource website at *www.explorasource.com/ educator* gives these correlations). The websites of many state departments of education also provide these correlations.

Exploring the Atlas Maps

Though the three excerpted maps in Figures i.1–i.3 represent many of the science learning goals for *Dig In!*, you will want to visit the AAAS Project 2061 website at *www.project2061.org* to download and explore other maps relevant to *Dig In!* lessons. The full *Atlas* maps also display strands of science learning for K–12; for reasons of space, the maps printed here only show K–8 science learning goals.

The maps suggest the progression of learning. A particular *Dig In!* lesson

may not be sufficient for students to become proficient with some of the basic or extended ideas in the map strand; checking the progress of your students along the way will help you see how to adapt instruction. Lessons may also touch on concepts outside of what the various science standards consider essential for basic science literacy. Therefore, you may decide to focus activities to achieve your core learning goals.

In addition to using the maps to plan instruction, you may wish to annotate maps with common student misconceptions, or accurate conceptions you can invoke to dispel these misconceptions. Motivating questions that have worked for you, and phenomena that illustrate points, may also find a place on your annotated maps.

Dig In! activities are interdisciplinary, covering art, geography, language arts, math, and social studies as well as science. You can annotate the *Atlas* maps with opportunities to achieve learning goals for other disciplines, or use the map design to trace your lesson plan structure for learning goals in the other disciplines.

National Science Teachers Association

Figure i.1. *Atlas of Science Literacy* map: Changes in the Earth's surface.

This map was adapted from *Atlas of Science Literacy* (AAAS 2001). For more information, or to order, to go *www.nsta.org/store.*

Map Key

BSL	concept from *Benchmarks for Science Literacy* (AAAS 1993)
code (e.g., 4F/2)	chapter, section, and number of corresponding Benchmark goal
SFAA	concept from *Science for All Americans* (AAAS 1993)
New Bench-mark	concept from a newly written benchmark
▦	concept covered in *Dig In!*

6–8

Thousands of layers of sedimentary rock confirm the long history of the changing surface of the Earth and the changing life forms whose remains are found in successive layers. The youngest layers are not always found on top, because of folding, breaking, and uplift of layers. 4C/5

Sedimentary rock buried deep enough may be reformed by pressure and heat, perhaps melting and recrystallizing into different kinds of rock. These reformed rock layers may be forced up again to become land surface and even mountains. Subsequently, this new rock too will erode. Rock bears evidence of the minerals, temperatures, and forces that created it. 4C/4

The interior of the Earth is hot. Heat flow and movement of material within the Earth cause earthquakes and volcanic eruptions and create mountains and ocean basins. 4C/1

The Earth first formed in a molten plate and then the surface cooled into solid rock. (New Benchmark)

Some changes in the Earth's surface are abrupt (such as earthquakes and volcanic eruptions) while other changes happen very slowly (such as uplift and wearing down of mountains). 4C/2

Vibrations in materials set up wavelike disturbances that spread away from the source. Sound and earthquake waves are examples. 4F/4

Sediments of sand and smaller particles (sometimes containing the remains of organisms) are gradually buried and are cemented together by dissolved minerals to form solid rock again. 4C/3

There are a variety of different land forms on the Earth's surface (such as coastlines, rivers, mountains, deltas, and canyons). (*BSL*, p.73)

The Earth's surface is shaped in part by the motion of water (including ice) and wind over very long times, which act to level mountain ranges. 4C/2

Rivers and glacial ice carry off soil and break down rock, eventually depositing the material in sediments or carrying it in solution to the sea. (*SFAA*, p.45)

Things on or near the Earth are pulled toward it by the Earth's gravity. 4B/1

Rock is composed of different combinations of minerals. Smaller rocks come from the breakage and weathering of bedrock and larger rocks. Soil is made partly from weathered rock, partly from plant remains—and also contains many living organisms. 4C/2

Chunks of rocks come in many shapes, from boulders to grains of sand and even smaller. 4C/1

Waves, wind, water, and ice shape and reshape the Earth's land surface by eroding rock and soil in some areas and depositing them in other areas, sometimes in seasonal layers. 4C/1

Things change in steady, repetitive, or irregular ways—or sometimes in more than one way at the same time. 11C/2

Change is something that happens to many things. 4C/2

Rocks and Sediments strand

Weathering and Erosion strand

3–5

How fast things move differs greatly. Some things are so slow that their journey takes a long time. 4F/2

Some changes are so slow or so fast that they are hard to see. 11C/4

Rates of Change strand

K–2

Earthquakes and Volcanos strand

Figure i.2. *Atlas of Science Literacy* map: Flow of matter in ecosystems.

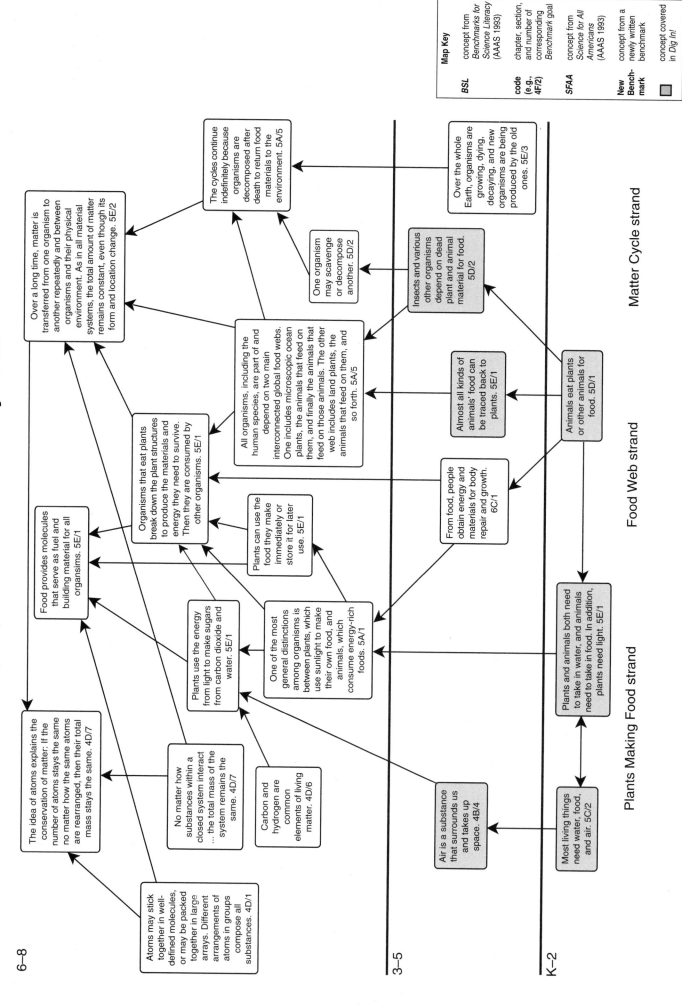

This map was adapted from *Atlas of Science Literacy* (AAAS 2001). For more information, or to order, to go *www.nsta.org/store*.

Figure i.3. Atlas of Science Literacy map: Agricultural technology.

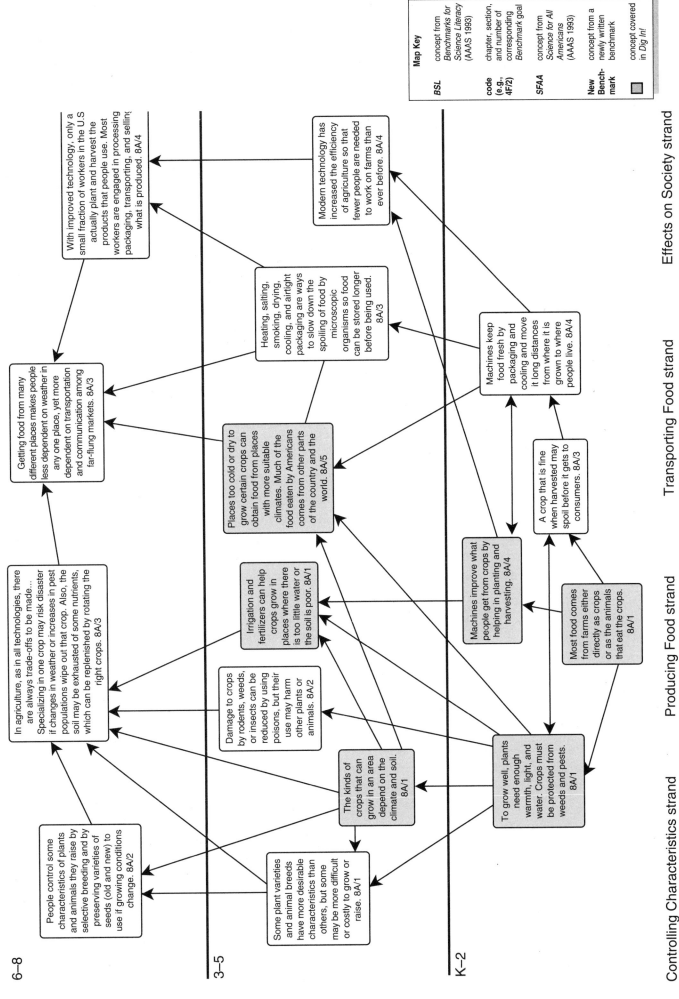

This map was adapted from *Atlas of Science Literacy* (AAAS 2001). For more information, or to order, to go *www.nsta.org/store.*

Lesson Assessment

Use the maps from *Atlas of Science Literacy* (Figures i.1–i.3) as an aid to your constructivist teaching methods. Allow students to recognize and integrate concepts— either never learned or incompletely remembered—into the big picture of why these concepts are useful to know.

Before you undertake any of the *Dig In!* activities, it is important to know whether your students have mastered the principles in the map that lead to their current grade level. You may, for example, be surprised to learn that some of your fourth-graders do not really understand that "Most food comes from farms either directly as crops or as the animals that eat the crops" (Benchmark 8A/1), a concept that, according to the *Atlas* map, should be mastered by grade two. Students may also have a mix of true and false understandings about what plants need in order to grow. It may be wise to ensure—perhaps by a class-developed concept map—that all students start *Dig In!* with the basic information needed to understand the concepts presented in these lessons.

The lessons in *Dig In!* use an inquiry approach to guide student understanding of the concept goals. Each lesson takes students through a learning cycle of making observations, exploring, making predictions, and testing their conclusions. Throughout each lesson, guided discussion, plus verbal and written presentation methods for students, provide a record of thinking and learning—showing you and your students what they understand, what is still fuzzy or missing, and whether students can now use what they know. Use these methods as formative assessments to guide your plans for how to use the *Dig In!* activities.

Each activity ends with an Evaluation section to assess your students' understanding of the concepts. The goal of any summative assessment is to determine whether students can apply their learning to new situations—to show you, and to show themselves, that they have a new tool for understanding. There are many ways to evaluate your students, and you should adapt your methods to the activity and to your class. For instance, Lesson 1 emphasizes ways to describe and characterize soil. An appropriate assessment method may be to have students draw the three types of soil and then fill in the blanks or use vocabulary words to label the drawings. Figure i.4 has an example of a rubric you might use to assess students for this lesson. You can adapt this rubric to match the Student Objectives of any of the lessons.

NSTA's *Science Educator's Guide to Assessment* has more guidelines on this topic (see the resources list in Appendix B).

Extensions

Each lesson concludes with a set of extensions for ways students can expand and test their learning. These extensions are applications of the concepts students have just investigated. You may wish to build a rubric for one or more of these extensions to share with your students, and use the extension as a summative assessment of your students' mastery of concepts.

Figure i.4. A sample assessment rubric for Lesson 1.

Activity	Excellent	Good	Fair
Description of dirt and soil	Students explain that dirt is a negative term and that soil refers to a valuable Earth material.	Students have difficulty describing the difference between dirt and soil.	Students cannot describe the difference between dirt and soil.
Comparison of clayey, silty, and sandy soil types	Students describe differences between size and texture of the three soil types.	Students describe only one difference between the three soil types.	Students cannot show any differences between the three soil types.
Identification of materials in soil	Students identify six or more materials in soil.	Students identify three to five materials in soil.	Students identify only one or two materials in soil.

Interdisciplinary Correlations

The activities in *Dig In! Hands-On Soil Investigations* are designed for elementary science classes, and can also be used to teach concepts in art, geography, language arts, mathematics, and social studies (see Figure i.5).

Figure i.5. Subjects covered by *Dig In!*

		Art	Geography	Language Arts	Mathematics	Science	Social Studies
Section I	Lesson 1: Soil Searching	•		•		•	
	Lesson 2: Scenic Soil	•		•		•	
	Lesson 3: Soil Supreme	•		•		•	
Section II	Lesson 4: Lofty, Level, and Lumpy	•	•	•		•	•
	Lesson 5: Life on the Land	•	•	•		•	•
	Lesson 6: Plant a Plant	•		•	•	•	
Section III	Lesson 7: Animal Apartments	•		•		•	
	Lesson 8: Living Links	•		•		•	
	Lesson 9: Watching Worms	•		•		•	
Section IV	Lesson 10: Wind + Water = Waste	•		•	•	•	•
	Lesson 11: Soil Scientists	•		•		•	•
	Lesson 12: An Outdoor Learning Center	•		•	•	•	•

National Science Teachers Association

Correlations with the *National Science Education Standards*

Dig In! is built upon the precepts of the *National Science Education Standards* for kindergarten to fourth grade. The science content standards outline what students should know, understand, and be able to do in the natural sciences over the course of K–12 education. Figure i.6 on the next page shows the correlation between those content standards and the concepts addressed in *Dig In!*

Figure i.6. Correlations with the *National Science Education Standards* for grades K–4.

Content Standard	Topic	Section I			Section II			Section III			Section IV		
		Lesson 1	Lesson 2	Lesson 3	Lesson 4	Lesson 5	Lesson 6	Lesson 7	Lesson 8	Lesson 9	Lesson 10	Lesson 11	Lesson 12
(A) Science As Inquiry	Ask a question about objects, organisms, and events in the environment	•	•				•	•	•	•	•		•
	Plan and conduct a simple investigation			•			•			•	•		•
	Employ simple equipment and tools to gather data and extend the senses	•	•				•	•			•		•
	Use data to construct a reasonable explanation						•			•	•		•
	Communicate investigations and explanations	•	•	•	•		•	•	•	•	•		

Figure i.6 continued

Content Standard	Topic	Section I			Section II			Section III			Section IV		
		Lesson 1	Lesson 2	Lesson 3	Lesson 4	Lesson 5	Lesson 6	Lesson 7	Lesson 8	Lesson 9	Lesson 10	Lesson 11	Lesson 12
(C) Life Science	The characteristics of organisms	•			•		•	•	•	•			•
	Organisms and their environment		•		•	•		•	•	•		•	•
(D) Earth and Space Science	Properties of Earth materials		•	•	•		•	•		•	•	•	•
	Changes in the Earth and sky		•							•	•	•	•
(F) Science in Personal and Social Perspectives	Types of resources					•	•	•	•	•	•	•	•
	Changes in environments				•	•		•	•	•	•	•	•
	Science and technology in local challenges				•	•					•	•	
(G) History and Nature of Science	Science as a human endeavor					•						•	•

sciLINKS

Dig In! Hands-On Soil Investigations brings you sciLINKS, a creative project from NSTA that blends the best of the two main educational "drivers"—textbooks and telecommunications—into a dynamic new educational tool for all children, their parents, and their teachers. This sciLINKS effort links specific textbook and supplemental resource locations with instructionally rich Internet resources. As you and your students use sciLINKS, you will find rich new pathways for learners, new opportunities for professional growth among teachers, and new modes of engagement for parents.

In this sciLINKed text, you will find an icon near several of the concepts you are studying. Under it, you will find the sciLINKS URL (www.sciLINKS.org) and a code. Go to the sciLINKS Web site, sign in, type the code from your text, and you will receive a list of URLs that are selected by science educators. Sites are chosen for accurate and age-appropriate content and good pedagogy. The underlying database changes constantly, eliminating dead or revised sites or simply replacing them with better selections. The ink may dry on the page, but the science it describes will always be fresh.

sciLINKS also ensures that the online content that teachers count on remains available for the life of this text. The sciLINKS search team regularly reviews the materials to which Dig In! points—revising the URLs as needed or replacing

webpages that have disappeared with new pages. When you send your students to sciLINKS to use a code from this text, you can always count on good content being available.

The site selection process involves four review stages:

1 A cadre of undergraduate science education majors searches the World Wide Web for interesting science resources. The undergraduates submit about 500 sites a week for consideration.

2 Packets of these webpages are organized and sent to teacher-web-watchers with expertise in given fields and grade levels. The teacher-webwatchers can also submit webpages they have found on their own. The teachers pick the jewels from this selection and correlate them to the National Science Education Standards. These pages are submitted to the sciLINKS database.

3 Scientists review these correlated sites for accuracy.

4 NSTA staff approves the webpages and edits the information for accuracy and consistent style.

Who pays for sciLINKS? sciLINKS is a free service for textbook and supplemental resource users, but obviously, someone must pay for it. Participating publishers pay a fee to NSTA for each book that contains sciLINKS. The program is also supported by a grant from the National Aeronautics and Space Administration.

National Science Teachers Association

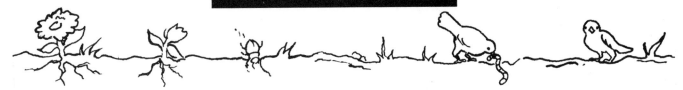

What Is Soil?

Overview

Soil is the solid material on Earth's surface that results from the interaction of weather and biological activities with the underlying geologic formation. Soil is produced from broken down rocks, organic matter (decayed animal and plant life), water, and air. Soil generally loosens from its parent material at a rate of one centimeter every 250 to 2,500 years.

Nearly 21,000 soil types are found in the United States. All soil types are made of varying amounts of three main components—silt, sand, and clay—and can therefore be classified as silty, sandy, or clayey soils. Many different colors can be present in soil, and depend on the minerals found in the parent material and on the chemical and biological reactions within the soil.

Each soil type is suited for a different use. Some soils, for example, can support the massive weight of buildings, shopping centers, airports, and highways. Some are best for crops or ranging land, some for wildlife habitat and forests. Soil scientists determine the capabilities of different soils based on texture, structure, depth, slope, organic matter, and chemical composition.

Soil is normally found in layers. Soil layers are distinguished by different colors, textures, and structures. Soil layers also have different amounts of plant and animal material (called "organic matter") and gravel.

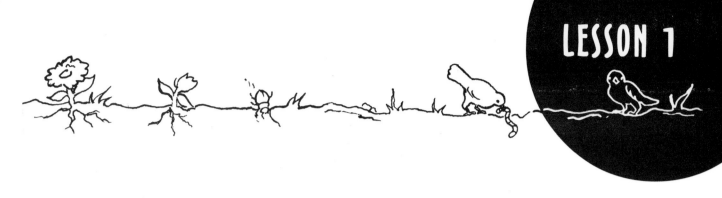

Soil Searching

Lesson Description

Students collect and handle samples of clayey, silty, and sandy soil.

Teacher Background

This lesson encourages students to think about the differences in *soil*. Soil is a naturally occurring mixture of *organic matter*, water, air, and minerals that forms on the surface of the land.

At first, most young learners make no distinction between soil and *dirt*. However, the differences should become clear with more careful thought. Dirt is soil that is out of place in the human world; for example, dust on the floor or mud on your shoes is often called dirt. Soil is the useful substance in which our food grows; the outermost solid surface of Earth that supports our cities, houses, and highways; and the medium that contains the minerals for plant and animal life. This thin layer of material may mean the difference between poverty and prosperity—even life and death—for all who inhabit the planet, since soil is the medium in which most of our food is grown.

There are three main components of soil: *clay*, *silt*, and *sand*. Clay is the smallest particle, with less than a 0.002-millimeter diameter. Silt particles are between 0.002 and 0.005 millimeters in diameter, and sand is the largest particle, ranging from 0.05 to 2.0 millimeters. Each soil has a characteristic *texture* that

Subjects

Art, Language Arts, Science

Time

Prep: 30 minutes
Activities: 1 ¾ hours
(not including Extensions)

Topic: soil
Go to: *www.scilinks.org*
Code: DIG01

Student Objectives

Students will be able to:

- differentiate between soil and dirt;
- recognize three basic soil components through sight and touch; and
- identify the materials in soil.

Materials

For the Class

- White glue
- Three index cards
- Clayey, silty, and sandy soils (see page x)
- Permanent marker
- Newspaper
- Two clear plastic jars with lids— e.g., tennis ball containers
- Small rocks
- Squeeze bottle
- Water
- Bucket
- Golf ball
- Softball
- Basketball

depends on its main component: *clayey* soils are fine, but may turn sticky and form a lump when wet, *silty* soils are smooth like flour but crumble when wet, and *sandy* soils are grittier than the others.

Students may find the following materials in soil: wood, rocks, roots, leaves, seeds, pods, stems, bark, grass, corn cobs, insects, insect eggs, and worms. Students will also find that soil's *pore spaces* contain air, and that at certain times soil also holds moisture.

Learning Cycle

Perception: 30 minutes

1   Ask students about the difference between soil and dirt. Help students understand that people think dirt is a nuisance. "Dirt" is a negative term while "soil" is a positive term for something useful. Dirt can be dust on the floor and mud on shoes; soil is the medium in which plants grow and animals live, a material without which people can't survive.

2   Demonstrate that soil contains air by filling a plastic jar half-full with soil. Slowly add water to approximately two centimeters from the top of the jar; air bubbles will rise as water displaces the air in the soil. Ask students why the air bubbles occur and guide them to the correct answer—that soil contains air. The air is contained in pore spaces in the soil.

3   Demonstrate that soil forms from the breakdown of rocks: put small clean rocks into a jar of water, then cover and shake vigorously. The water should turn cloudy as soil particles loosen from the rocks. Ask students to observe the jar carefully and then discuss their observations.

4 Clean up by emptying the two jars into the bucket, than dispose of the wastewater outside rather than in the sink or trash.

Exploration: 30 minutes

Prep Cover a demonstration table with newspaper. Spread a dollop of glue on each of the three index cards. Sprinkle one type of soil on the glue on each of the cards and use a marker to label samples "silty soil," "sandy soil," and "clayey soil." Allow index cards to dry.

Prepare student work areas by covering tables with newspaper. Each work area should have dry paper towels, damp paper towels, a magnifier, a spoon, and three labeled cups each filled halfway with the different soils.

1 Distribute hand magnifiers. Demonstrate how to use the magnifiers and allow students to examine clothing, newsprint, or their fingers for a few minutes.

2 Have students put on their smocks or shirts, then dump the silty soil onto a dry paper towel and examine the soil with magnifiers.

3 Ask students what they see in the soil, and list the discoveries on the board. Help younger students distinguish pieces of rocks, plant material, and twigs. By touching the soil, students may discover that their soil samples contain moisture.

4 Discuss the texture of the silty soil. Ask students to suggest words that describe how the soil feels on their fingers.

Materials Cont'd.

For Each Student Group
- Clayey, silty, and sandy soils (see page x)
- Three clear plastic cups
- Spoon
- Small plastic hand magnifier (approximately 5x magnification)
- Paper towels
- Resealable plastic sandwich bag
- Smocks or old shirts

5 Demonstrate how to use a spoon to scrape the silty soil into a neat pile on its paper towel and then ask students to do the same.

6 Tell your students that they just looked at one type of soil. Ask students to predict what the other two types of soil will contain. You may wish to list predictions on the board.

7 Repeat the discovery process with sandy soil and clayey soil.

8 Guide students to figure out which soil particle—clay, silt, or sand—is largest based on their discoveries. (Answer: clay is the smallest, sand is the largest.)

9 Model the size difference between particles using three types of balls: if a clay particle is represented by a golf ball, then a silt particle would be the size of a softball and a sand particle would be the size of a basketball.

10 Clean up, saving the materials on the demonstration table and work areas for the Application section. Make sure students wash their hands.

Application: 30 minutes

 Use the demonstration table and work-area setup from the Exploration section.

1 Ask students to predict what might happen to each pile of soil as drops of water are added. List student predictions on the board.

2 Add a few drops of water from a squeeze bottle to students' soil piles, and ask students to observe and describe what happens.

3 Have students make three soil balls using soil from each pile. Students should wipe their hands on damp paper towels between handling different soils.

4 Ask students to describe what happens when they make soil balls. Students should discover that wet clayey soil forms a lump, wet silty soil crumbles easily, and wet sandy soil runs through their fingers. Discuss observations and write results on the board. Explain that soil's reaction with water is a model for what happens when rain falls on different kinds of soil in our yards, gardens, and fields.

5 Clean the demonstration table and work areas. Keep the piles of dry, unused soil for Lessons 2, 3, 6, 9, and 10, which also require soil. Collect wet and mixed soil in the bucket, then dispose of the waste material outside.

Evaluation: 15 minutes

Students should be able to describe the difference between dirt and soil, and compare silty, sandy, and clayey soil. They also should be able to identify some of the materials in soil based on their observations and discussions. Younger students can draw and color the three components of soil, while older students might label drawings of soil and choose the appropriate vocabulary words from a list that you provide (e.g., *dirt, soil, silty soil, sandy soil, clayey soil, pore space, rock, organic matter, twig, color, texture*). Figure i.4 on page xvii (in the introduction to this book) shows an example of a rubric you might use to assess your students.

Extensions: 30 minutes each

- Investigate settling rates of soil. Fill three jars three-quarters full of water and add a few drops of Calgon® bath gel or biodegradable liquid dish soap to each jar (soap speeds up settling). Add silty soil to the first jar, sandy soil to the second, and clayey soil to the third. Label the jars with tape and a permanent marker. Ask students to predict what will happen when the jars are shaken. Cover, shake the jars, and allow the soil to settle. Were students' predictions correct?

- Make mud pies. Students should first guess which soil would make the best mud pie, based on what they have learned. (Answer: clayey soil, because it sticks together when wet more than the other soils.) Demonstrate how to make a mud pie with clayey soil and water. Have students make a pie, then press a leaf into it to make a pattern or decorate the pie with other plant material. Explain that in nature, leaves and twigs land on the soil, eventually break into tiny pieces, and become a part of the soil, making the soil loose and dark.

- Take soil samples. Bring the class outside and dig a small hole in the ground, several centimeters deep. Have students feel the soil samples for moisture. If possible, repeat this activity throughout the year, and relate the soil moisture to precipitation, temperature, and season.

Scenic Soil

Lesson Description

Students discover how soils are formed, then paint pictures to show soil colors.

Teacher Background

Soils form over millions of years from *parent material* that is broken down by *weathering* from wind, water, temperature, chemical changes, and living organisms. Over time, glaciers move over the land and grind rocks together, rubbing off particles of all sizes. By day rocks are warmed by the Sun and expand, while at night the rocks cool and contract. Over time, enough expansion and contraction cause rock particles to chip off. Change in water temperature also contributes to soil formation—in cold temperatures, water in the cracks of rocks freezes and expands, causing the rocks to break into smaller pieces. Growing plant roots can also split a rock, contributing to soil formation from parent rock material. When the rock particles from these various sources combine with living microorganisms, air, moisture, and *organic matter* from *decomposing* plants and animals, soil is formed.

A student's first impression when looking at bare soil may be of its color. You may be familiar with the Painted Desert in Arizona, red desert soils in California, Arizona, and Nevada, or the dark, fertile soils of the Plains states. Many early human cultures used earth materials (such as the soil mentioned above) to color ceramics, cosmetics, and paints.

Subjects

Art, Language Arts, Science

Time

Prep: 30 minutes
Activities: 1 ¾–2 hours
(not including Extensions)

Student Objectives

Students will be able to:

- explain how soil is made;
- recognize that soil is composed of different colors; and
- design a soil painting using those different colors.

Materials

For the Class

- Newspaper
- Index cards
- White glue
- Water
- Clear plastic or glass jars with lids (any size)
- Spoon
- Bucket

For Each Student Group

- Resealable plastic sandwich bag
- Spoon
- Paper towels
- Small plastic hand magnifier (approximately 5x magnification)
- Plastic jar with lid (approximately 0.35 Liters)

Soil color and development are parts of weathering. For example, as rocks containing iron are exposed to air, the elements turn yellow or red. Organic matter in soil decomposes into fertile black *humus*. The element manganese forms black mineral deposits in soil, while quartz can create white layers in the soil. Soils deep below the surface become lighter, redder, or more yellow. Soils influenced by water are gray or have gray spots, depending on how long the water is present and where it is located in the soil. Climate, physical geography, and geology influence the rates and conditions under which these chemical reactions occur and color the soil.

Learning Cycle

Perception: 30 minutes

Prep Set up the demonstration table and student work areas as in Lesson 1. Each work area should have dry and damp paper towels, and a magnifier.

1   Take the class to the school yard. Give each student group a spoon and small plastic bag.

2   Ask each student group to dig up one spoonful of soil and put it in the plastic bag. For best results, dig loose and crumbly soil. If inclement weather or frozen ground makes this impossible, bring in your own soil samples (see page x for suggestions on obtaining soil).

3   Back in the classroom, ask students to match the school-yard soil sample with one of the larger samples on the index cards. Pass around plastic bags and help students distinguish differences in color and texture.

4 On the demonstration table, sort soil samples into piles according to color. Use index cards and the marker to label the location of each soil. If the colors are too similar, add samples from other locations such as local parks and farms.

5 Have students examine the new samples with magnifiers at their work areas. Students may discover that their school-yard soil samples contain plant and animal matter, pebbles, and moisture. Compare the discoveries in the soil samples.

6 Ask students to observe changes in soil color as their samples dry. Students should become aware that soils are lighter when dry than when wet.

7 Clean up the demonstration table and work areas. Save the soil samples for the Exploration section.

Exploration: 30 minutes

Prep Cover student work areas with newspaper. Each work area should have a jar, a bag with a school-yard soil sample (from the Perception section), rocks, grass, leaves, twigs, and a bottle of water.

1 Seat students at work areas and have them put on smocks or old shirts.

2 Have students create new soil by mixing equal parts of soil from the demonstration table with grass, bits of leaves, twigs, and drops of water. Students can rub rocks together to add minerals to their recipe. (Pebbles and rock chips will weather faster than larger rocks.) Have students put their mixtures into jars, tightly screw on the lid, and gently shake. Students in each group should write their names on tape and label their jar.

Materials Cont'd.

- Several soft, crumbly sedimentary rocks—e.g., shale, limestone, or sandstone
- Grass clippings
- Dried leaves
- Twigs
- Squeeze bottle
- Water
- White tape
- Permanent marker
- White watercolor paper or posterboard (approximately 21.5 x 27.9 centimeters/8.5 x 11 inches)
- Pencil
- Ruler or straightedge
- Three to five clear plastic cups
- Egg carton
- Watercolor brush
- Jar without lid for water
- Toothpicks
- Smock or old shirt
- Student Handout 2A (optional)

3 Gather the jars and ask students to name any differences among the jars' color and texture. You may wish to record student observations on the board. Review what makes the colors and textures.

4 Once a week for the next few months, have students add water and then gently shake their jars. Students can draw and label pictures of their jars as the leaves and clippings become part of the soil.

5 After your class has finished observation of the jars, empty the two jars into the bucket, then dispose of the waste material outside.

Application: 30 minutes

Prep Mix white "paint" by combining equal parts of white glue and water in jars, then seal the jars. (Experiment with painting designs to determine how much paint your students will need.) Cover student work areas with newspaper. Each work area should have one or two pieces of watercolor paper or posterboard, a ruler, a pencil, a watercolor brush, an egg carton, a jar with water, toothpicks, a spoon, and several plastic cups with different colored soil.

1 At their work areas, older students can use a pencil and ruler to trace or draw a design using basic shapes on a piece of watercolor paper or posterboard. The design could be made of overlapping circles, squares, rectangles, and triangles (see Figure 2.1). For younger learners, prepare the colored paints beforehand and label jars with the names of the colors. Focus on painting using three basic colors, and then have students give names to both the colors and the soil.

National Science Teachers Association

Figure 2.1. A design for soil painting.

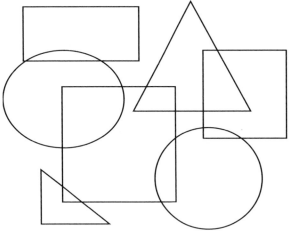

2 Ask students to put on their smocks and then spoon different colored soils into spaces in the egg cartons. Students can mix soil types to create different colors, but no more than six of the twelve spaces should contain soil.

3 Walk around the class and for each student group, ladle two spoonfuls of white paint into each egg carton space containing soil.

4 Students can mix each soil paint with a different toothpick, then paint their designs.

5 After the paintings dry, hang all the pictures to show the variety of colors found in soil.

6 Clean up work areas. Keep the piles of dry, unused soil for Lessons 3, 6, 9, and 10, which also require soil. Collect wet and mixed soil in the bucket, then dispose of the material outside. Dispose of waste material that contains glue in a trashcan.

Evaluation: 15–30 minutes

Ask students to present their soil paintings and use the paintings to describe the different soil colors and how soil forms.

Extensions: 30 minutes each

- Experiment with soil colors in a way that may be easier for young students: distribute copies of Student Handout 2A. Instruct students to color the flower and roots, use a cotton swab to carefully smear glue in and around the roots, and then sprinkle soil on all the glue areas. Allow the pictures to dry before gently shaking off excess soil.

- Get moist clay from an art teacher, pottery shop, or crafts store. Remind students that clay is one of the soil components discussed. Have students push their thumbs into the clay, then press their thumbprint onto paper. Repeat this process several times. Have students add character to these thumbprints by drawing lines to make faces, limbs, tails, and feet, or stems and leaves.

- Ask students to bring to class soil samples from their yards or local parks. Compare the soil textures, colors, and observations.

Name: _____

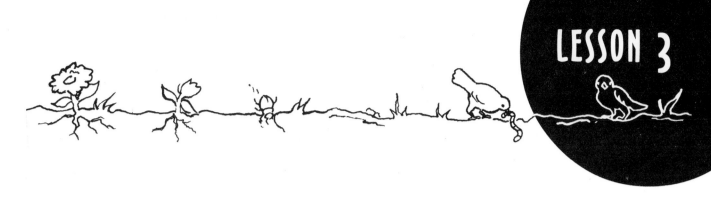

Soil Supreme

Lesson Description

Students investigate soil layers with three models: a jar of water and Earth materials, illustrations on paper, and edible materials.

Teacher Background

This lesson reveals a world that is usually hidden from view. It usually takes deep digging or a field trip to a road cut to discover Earth's three predominant soil layers. But through fun activities and a model of different soil types, students will learn how soil forms layers and about the world of living things in the soil.

Soils have different layers. The layers are based on soil color, texture, and structure, and on the amount of organic matter and gravel present in the soil. You can differentiate between the layers because of these variations. Differences in layers based on color are the easiest to identify. In areas where the soil has been disturbed—such as in agricultural areas, building sites, or places with severe erosion—there may be only one or two naturally occurring layers present.

The uppermost layer of soil is usually the most productive part. It is characteristically darker and looser than the lower layers. Fallen twigs and leaves, dead grass and insects, rotting tree trunks, and other decaying organic matter enrich the top part, giving living animals and plants the nutrients they need to survive.

Subjects

Art, Language Arts, Science

Time

Prep: 30 minutes
Activities: 1 ½–2 hours
(not including Extensions)

SC*L*INKS
THE WORLD'S A CLICK AWAY

Topic: soil layers
Go to: *www.scilinks.org*
Code: DIG03

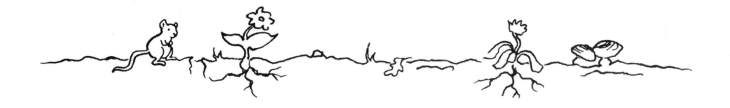

Student Objectives

Students will be able to:

- explain the differences between topsoil, subsoil, and parent material; and
- construct models of the three soil layers.

The middle layer of soil is often lighter, denser, and grittier than the top part. This middle layer is usually low in organic matter and frequently contains small stones and pebbles. The next part—a transition area between the soil layers and rock—contains disintegrating rock particles, but has relatively little organic life. The deepest layer is the *parent material*, or *bedrock*, from which much of the soil originally formed.

Cross sections and *models* will help students visualize the soil layers.

Materials

For the Class

- Plastic jar (at least 1 Liter) with lid—e.g., tennis ball container
- Silty, sandy, and clayey soil samples from Lessons 1 and 2 (or see page x)
- Sand
- Pebbles
- Twigs
- Grass clippings
- Dried leaves
- Biodegradable liquid dish soap
- Spoon
- Bucket
- Construction paper squares (approximately 10 x 10 centimeters)—black, gray, dark brown, medium brown, light brown, and orange
- Poster paper or bulletin board
- Tape, glue, or pushpins
- Cake pan (33 x 23 centimeters/13 x 9 inches)

Learning Cycle

Perception: 30 minutes

Prep Ask the local Natural Resources Conservation Service office for a county soil survey, which should give you information to adapt color choices to soil colors of your area (see the resources list in Appendix B).

Fill a large plastic jar one-quarter full of soil, pebbles, twigs, and plant matter. Fill the jar three-quarters to the top with water and add a few drops of dish soap. If desired, create charts on which students can record their predictions and observations during the activity.

1 Make sure the jar lid is secure. Shake the jar vigorously and let stand.

2 Ask students what they observe and discuss what the resulting layers are made of. You may wish to have students draw a series of sketches, or complete prediction/observation charts that you created beforehand.

National Science Teachers Association

3 Pour off the water without disturbing the layers. Use a spoon to sample each layer to show students that the soil components separated based on size. The densest particles—sand and pebbles—sink to the bottom of the jar. Silt forms the middle layer, clay forms the top layer and dead leaves, sticks, and other organic matter float on top of the water.

4 Make "before" and "after" sketches of the jar on the board, and have students help you label the sketches.

5 Distribute Student Handout 3A. Explain that the drawing is a cross section, a side-view of what the soil looks like in the ground. Ask students to draw on the empty jar illustration the things that are found in the layers, then to label each layer on the line provided. You may wish to add vocabulary word choices to this handout so younger students can choose the appropriate words to label the illustration.

6 Clean up by emptying the jar into the bucket, than dispose of the waste material outside.

Exploration: 30 minutes

1 Divide students into three groups and name them: Topsoil, Subsoil, and Parent Material.

2 Guide students to conclude what color their group should be, based upon their discoveries about soil. (Answers: the top soil layer is usually black, dark brown, or gray. The middle layer is often medium brown or light brown. Parent material is usually light brown, tan, or white. However, soil colors depend on what part of the country you live in.)

3 Ask each student to pick a piece of colored construction paper based on his or her group. Have groups gather and draw on the construction

Materials Cont'd.

- Crushed vanilla wafers (340 grams/12 ounces)
- Soft chocolate ice cream or chocolate pudding (2.20 Liters/ 2 quarts)
- Crushed dark chocolate cookies (340 grams/12 ounces)
- Six gummy worms (optional)

For Each Student Group

- Crayons, colored pencils, or markers
- Clear plastic cup
- Crushed vanilla wafers (21.26 grams/0.75 ounces, or approximately 6 cookies)
- Soft chocolate ice cream or chocolate pudding (0.275 Liters/ 0.5 cups)
- Crushed dark chocolate cookies (21.26 grams/0.75 ounces, or approximately 6 cookies)
- Gummy worm (optional)
- Spoons
- Napkins
- Student Handouts 3A and 3B

paper organisms and objects found in their particular soil layers.

4   Guide groups in the use of pushpins or tape to arrange their colored squares in correct layers on the bulletin board or poster paper. Discuss the papers' resemblance to real soil layers.

Application: 15–30 minutes

Prep Prepare one cake pan and clear plastic cups of "Soil Supreme." In the pan and each cup, first put down a layer of crushed vanilla wafers—this represents the parent material. Then create a layer of chocolate ice cream or pudding, which represents the middle part of soil. Pour on crushed dark chocolate cookies for the top layer. Add gummy worms to the top layer if desired. Freeze for at least two hours before serving. (Recipe makes approximately 16 servings.)

1   Cut a piece of Soil Supreme and discuss the layers of this cross section with students. Help students understand that the cake is a model of the real world.

2   Hand out pieces of Soil Supreme and enjoy. (Remember to eat in a clean classroom rather than in a science laboratory, and be mindful of any food allergies.)

3   Clean eating areas.

Evaluation: 15–30 minutes

Students should be able to name and briefly describe the three main soil layers. Students can color and label each layer of soil on Student Handout 3B. You may wish to provide younger students with a list of vocabulary words to help them label layers. You might

also ask students to draw things found within the soil layers.

Extensions: 30 minutes each

- To help students understand the complicated soil horizons, find a site where soil layers are exposed. Take the class on a field trip to a quarry, a site of extensive erosion, a streambank, or an excavation. Look at the soil and measure the depth of the horizons. To enhance students' observational skills, ask them to look for and record changes in soil color, rocks, wood fragments, roots, and other material in soil. Ask students why the layers are different or the same. (Answer: the characteristics of each layer are due to local geologic history, climate, biological activity, land use, and amount of time that the soil has been developing and eroding.) If a field trip is not possible, bring in photographs that show these soil horizon characteristics.

- Look for examples of cross sections in the school yard or in neighboring areas. Bring in pictures of cross sections, including a scientific illustration.

- Depending on the age of your students, prepare Soil Supreme in class and then eat it during the next class period. This will give your students a good introduction to measurements. Students can participate in crushing the cookies—put the cookies in sealed plastic sandwich bags and then have students crush the cookies with rolling pins. Letting your students measure the edible materials for Soil Supreme will give students a good introduction to measurements.

Name: _____

① ②

② ②

③ ③

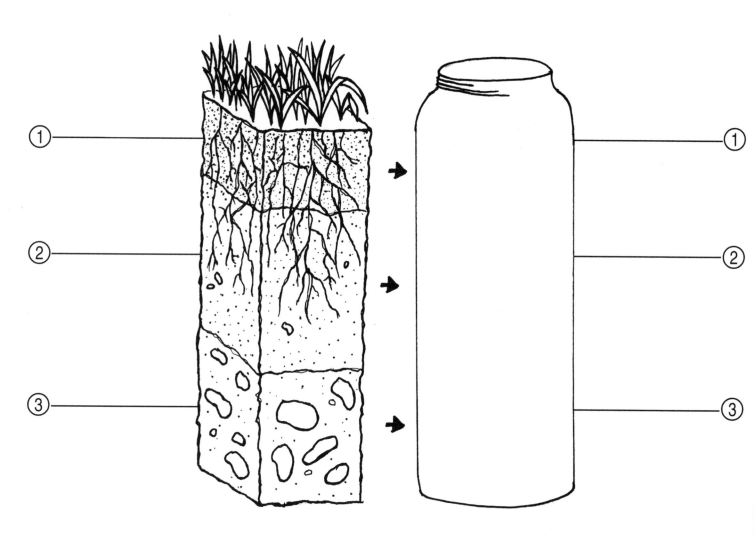

Name: _____

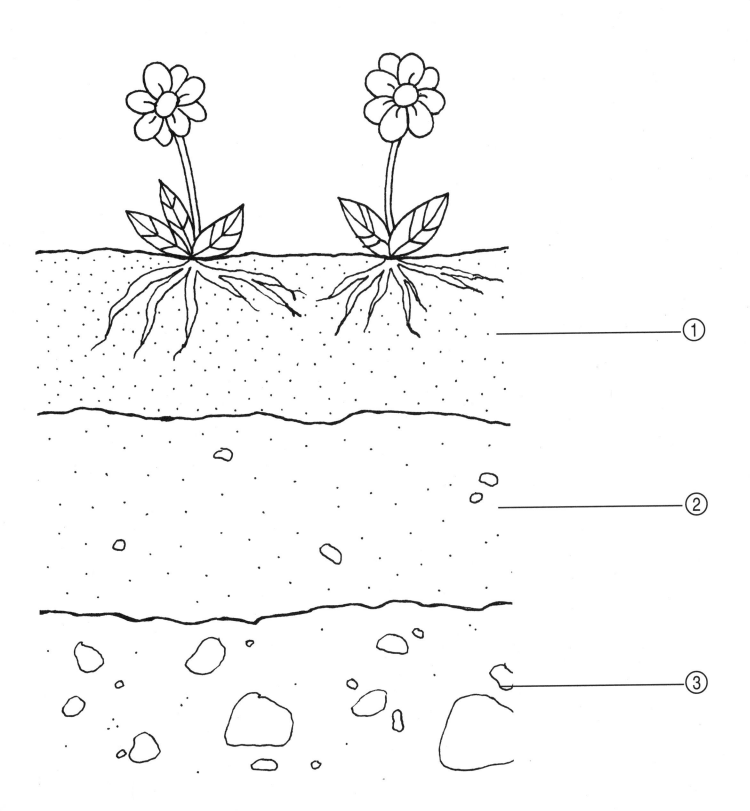

Who Uses Soil?

Overview

The study of geography suggests relationships among elements of the natural world. Moving water, weather, erosion, and gravity resculpt the land. Over the years, glaciers, volcanoes, fires, hurricanes, and earthquakes have also reshaped the Earth's surface, moving materials from one place to another. Two natural forces are continuously at work on the landscape—the forces of destruction and the forces of construction. In locations of destruction, the material that is lost through natural events must go somewhere else, and landforms are constructed. The landscape is continually developing.

When humans work the land, for agriculture, home building, and road construction, we change the landscape to fit our purposes. Land use involves manipulating the landscape and its components—the

soil, rocks, and vegetation. Humans change the land but differently than how nature changes the land.

National Science Teachers Association

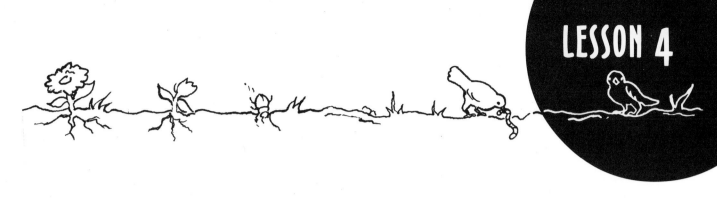

Lofty, Level, and Lumpy

Lesson Description

Students learn about habitats and the characteristic animals and plants of those habitats.

Teacher Background

There are five predominant *habitats* of the continental United States: *deserts*, *forests*, *mountains*, *prairies*, and *wetlands*. Each habitat supports a unique community of plants and animals, and has characteristic *landforms* and soil types.

Desert habitats are dry, hot, and have sparse vegetation. Desert animals include lizards, snakes, small birds, small mammals, and other animals adapted to dry, hot climates. Cacti and small wildflowers are typically found in deserts. Second, forests receive a medium amount of precipitation and are more temperate than deserts. A forest habitat is generally covered with dense vegetation, including trees, shrubs, and woody plants, that shade the land from the Sun. Many small and large mammals, birds, reptiles, amphibians, and insects can be found in the forest. The third habitat, mountains, are high-elevation areas often with rocky soil and steep slopes. Typical mountain dwellers include a variety of birds, small mammals, reptiles,

Subjects

Art, Geography, Language Arts, Science, Social Studies

Time

Prep: 30 minutes
Activities: 1 ¾ hours
(not including Extensions)

SCi LINKS®
THE WORLD'S A CLICK AWAY

Topic: habitats
Go to: *www.scilinks.org*
Code: DIG04

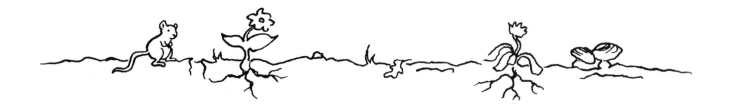

Student Objectives

Students will be able to:

- define habitats and landforms;
- compare habitats; and
- recognize that soils sustain different plant and animal life.

and insects. Trees can be found on mountains up to a certain elevation; only short woody plants and grasses grow at higher altitudes. Fourth, prairies are characterized by rolling hills or level land with a medium amount of rainfall. Prairie animals include birds, small mammals, reptiles, amphibians, and insects. Prairies normally consist of tall and short grasses, wildflowers, and few trees. The last main habitat, wetlands, are marshy, humid areas. Wetlands animals—such as waterfowl, wading birds, fish, and amphibians—live in or near the water. Wetland plants—such as cattails and mangrove trees—are also adapted to the wet environment.

Students will discover the properties of each habitat, describe what each region is like, and take an imaginary vacation in the region of their choice. A particular focus of this activity is differences in plant and animal life in the five habitats, and the lesson will demonstrate that plants and animals have features that help them live in different habitats.

Materials

For the Class

- United States or world map
- Pictures of deserts, forests, mountains, prairies, wetlands, animals, and plants
- Index cards
- Glue
- Scissors
- Poster paper
- Marker
- Tape

Learning Cycle

Perception: 15 minutes

Prep Photocopy or cut out pictures of deserts, forests, mountains, prairies, and wetlands, and familiar plants and animals from textbooks, encyclopedias, calendars, and magazines such as *National Geographic*, *Outside*, *Audubon*, and *Sierra Club*. Other good sources of habitat pictures are the World Wide Web and educational CD-ROMs. Include pictures of your local area. Paste the pictures onto index cards.

1 Help students understand what a habitat is.

National Science Teachers Association

2 Discuss the five main habitats of the United States. Use the pictures to discuss traits about habitats; for example, the desert is dry and hot, the forest is shady and has a lot of trees, the prairie is sunny and is covered with grass, the mountains are tall and rocky, and the wetlands are wet, humid, and swampy.

3 Ask students to name and describe the local habitat, including characteristic animals, plants, landforms, and soil types.

4 Make a summary chart, concept map, or word web with words and phrases about your local habitat.

Materials Cont'd.

For Each Student Group
- Tape
- Marker
- Index cards from Perception section
- Drawing paper
- Crayons, colored pencils, or markers
- Scissors
- Glue
- Student Handouts 4A–4G

Exploration: 45 minutes

Prep Create a blank concept map for each habitat (desert, forest, mountain, prairie, and wetlands) on separate pieces of poster paper. Each concept map should have space to list animals, plants, soil type, temperature, climate, landforms, and locations. Tape these maps on the wall.

Use the index cards from the Perception section, and separate the cards by habitat. Many of the animals and plants can fit into separate habitats, but each habitat should have an equal number of cards.

1 Break students into five groups and assign each group to a habitat concept map. Distribute to each group a marker, index cards, and tape. Choose one captain, one recorder, one speaker, and at least one traveler for each group.

2 The captain decides how to distribute the group's index cards and tapes them on to the concept map. For instance, the captain may decide that a card with a mountain belongs near the word "landform"

on the concept map, and a card with a deer belongs near the word "animal."

3   Guide travelers one group at a time, to the world or country map to determine the locations of their habitats. Travelers should also brainstorm animals, plants, soil types, temperatures, climates, and locations for the group's habitat.

4   The recorder writes the travelers' observations on the concept map.

5   Once the groups have completed their concept maps, the speakers will take the class on imaginary vacations to the habitat by presenting their maps. An example of a speaker's tour is: "I'll be your tour guide today in the desert habitat. We will be visiting the Southern Arizona area where you will see animals such as jackrabbits, mice, lizards, and rattlesnakes, plants such as cacti, and sandy soils. You won't see many people since the desert does not have a lot of inhabitants. The temperature is hot and dry, so be sure to dress in shorts and a t-shirt, but wear sunscreen because the Sun is strong. Bring a jacket since the nights are cool."

Application: 30 minutes

Prep   Make large photocopies of Student Handouts 4A–E, which represent the habitats your class has discussed.

1   Distribute one large habitat picture to each group. Also distribute to each group one copy (actual size) of each of the three pages in Student Handouts 4F and the two pages in 4G, which depict animals and plants. (Note that the drawings are not to scale.)

2   Assign two or three students in each group to color in the habitat picture.

3 Meanwhile, other group members cut out the animals and plants from Handouts 4F and 4G that belong in their habitat. Students then color in those animals and plants and paste them in an appropriate space on the habitat picture. If you are working with younger learners, you may wish to draw dotted outlines in which to paste each animal and plant picture on the habitat. Draw a line under the dotted outlines where students can label the pictures.

4 You can leave out the habitats that are not appropriate for your region. If your region is very different than any of the ones pictured on the handouts, invite groups to compare the habitat pictured with the local region, or draw their own pictures.

5 Ask students to label each animal and plant, and give the picture a title that includes the name of the habitat. You may wish to provide younger students with a vocabulary list to use when labeling the animals, plants, and habitats.

6 Many of the animals and plants can be found in more than one habitat. Discuss this with students so they understand that there are several possible correct pictures to create.

Evaluation: 15 minutes

Students should be able to explain what a habitat is, and describe how the local habitat differs—or is the same as—the habitat covered by their group. Students should be able to explain the clues they used to determine why each animal and plant from the handouts belonged in their habitat. Ask each student to share one fact they learned about the local habitat that they didn't know before this lesson.

Extensions: 30 minutes each

- Students may vote on their favorite habitat and make a model of it in the classroom, or decorate a bulletin board. For example, to create a "mountain" environment, students may draw mountains, trees, a lake or stream, rocks, and birds, or use family pictures from a trip to the mountains to share with class members.

- Ask students to name and locate on the map some of the places they have traveled. Have students discuss what those places were like. Guide the discussion to include the types of habitats and descriptions of the land or soil.

- If a student's family has photographic slides from a trip, ask parents to show the slides to class, and discuss the habitats, plants, and animals shown.

- To integrate the concepts to language arts and reading, have your students write stories or poems about local habitats.

- Investigate place names. You might focus on multicultural place names in the United States. For example, have students brainstorm Native American names given to places in New England (e.g., "Connecticut," "Massachusetts," and "Passaic River") or Spanish names in the Southwest (e.g., "Rio Grande River" and "Los Angeles"). Guide your students to consider what place names reveal about a location; for instance, think about the physical differences between Crater Lake, Lake Superior, and Great Salt Lake.

Name: _____

Name: _____

Name: _____

Name: _____

Name: _____

Name: _____

Name: _____

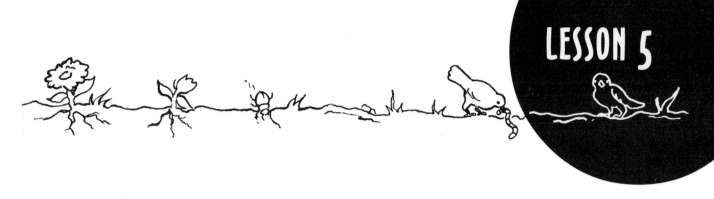

Life on the Land

Lesson Description

Students consider the other important uses for land besides food production.

Teacher Background

This lesson highlights wise *land use*. Land use refers to the ways in which private and public space is used. For example, land use can be residential, commercial, governmental, agricultural, or recreational. People often think of soil only as the substance in which we grow our food, but it has other important uses for humans.

One-third of Earth's surface is land, and only a portion of that land is suitable for human habitation. How we use this precious *natural resource* affects the health and happiness of all people. The factors that most influence land use are the physical properties of the land—the soil, slope, and water availability. Competition and other social and economic factors also come into play when communities decide how to use private and public lands. Often, a community's *planning commission* must make difficult decisions about the restrictions that apply to various land uses.

Encourage your students to consider land availability, soil productivity, locality, and safety during your

Subjects

Art, Geography, Language Arts, Science, Social Studies

Time

Prep: 15 minutes
Activities: 2 ½–2 ¾ hours
(not including Extensions)

Student Objectives

Students will be able to:

- give examples of humans' use of the land;
- demonstrate an awareness of the trade-offs involved in land use; and
- illustrate how a planning commission might develop a town.

Materials

For the Class

- Apple
- Small knife
- Vegetable peeler
- Stapler
- Chart or poster paper
- Paper towels or toilet tissue

For Each Student Group

- Several paper bags of different sizes
- Several crayons or markers
- Colored construction paper
- Tape
- Newspaper or used white paper

discussions about land use. For example, the buildings in a crowded, urban environment will differ from those in an open, rural environment.

Learning Cycle

Perception: 30 minutes

1 Ask students to think about where their food comes from. Is their food grown nearby or far away from their home? (Answer: most food is probably grown or raised in other parts of the country, transported, and then purchased at local grocery stores.)

2 If applicable, ask students to brainstorm foods grown locally.

3 Demonstrate the limited amount of land available for growing food with the model in Figure 5.1. Discuss what will happen to Earth if we lose such a precious resource as land. Relate the idea of land use to students' lives.

4 What is soil used for besides growing food crops? If applicable, discuss students' yards and gardens—do the gardens take up more space than the homes?

5 Brainstorm other uses of soil or land and list ideas on the board or poster paper. Examples of land use are:

❏ surfaces for homes, schools, hospitals, businesses, and airports

❏ surfaces for roads and highways

❏ surfaces for mining coal, ore, gravel, and minerals

National Science Teachers Association

- surfaces for parks, recreation, and wildlife
- land for growing food and fiber crops
- land for raising animals for food

Figure 5.1. Only a small portion of land is capable of producing food.

1 Imagine the Earth as an apple.

2 Cut it into fourths. Only one part is land—the rest is water. Set aside the three sections that represent water.

3 Cut the land section in half. One part represents land that is mountains, deserts, or covered with ice. Set this part aside.

4 Cut the other livable area into fourths. Three of these are too rocky, wet, hot, infertile, or covered with roads and cities to grow food. Set these three aside.

5 There is now only ¹/₃₂ of a slice of apple remaining. Peel the skin from this tiny piece.

6 The skin represents the soil on which the food is grown that must feed all the people on Earth.

6 Explain to students that sometimes there isn't enough land for all uses and people must decide what is most important. Describe town planning commissions—people who decide how to use the land.

Exploration: 45 minutes

1 Ask students to think about land uses they consider most important. How is land being used in your community? Record ideas on the board as students discuss important land uses.

2 Divide the class into small groups and assign two specific uses of land (e.g., home, fire station, school, park, store, gas station, city hall, hospital, power plant, and vegetable farm) to each group. Groups should spend about 10 minutes to determine how much space is needed for that particular use, if the institution should be in a business or residential area, and how much green space would be left.

3 Each group should present its plan for land use to the class, which acts as the town planning commission. The planning commission should vote on whether to approve the group's plan. The planning commission may refuse a request to build if members feel that a building will be built in an inappropriate part of town.

4 As a class, discuss how much space each building, farm, park, etc. needs. Have the class reach an agreement on how to plan the town. Help students understand that land space is important for many reasons besides growing food.

5 Have groups draw and label pictures of the planned town, or draw one large map on poster paper.

National Science Teachers Association

Application: 1 hour

Prep Clear a space in the classroom about 2.5-by-2.5 meters so that students can develop an ideal city that incorporates varying uses of the land.

1 Divide students into small groups from the Exploration section.

2 Pass out different size paper bags. Have groups decorate the bags like buildings using markers and construction paper, then stuff the bags with crumpled newspaper. You can staple the students' bags at the top to make buildings.

3 Use the maps to arrange the town. Set out the paper-bag buildings along strips of paper—toilet tissue or paper towel—that represent streets. Suggest adding a garbage dump, power plant, and parks, and have the class vote on where these buildings and areas should be built.

4 As students arrange their city, guide the discussion to past and future uses of land in your own town.

Evaluation: 15–30 minutes

Observe students as they work and discuss land use within their groups. At the end of the activity, each student should be able to identify at least three uses of land and explain why those uses are important. For example, students might list houses or apartment buildings as places for people to live, businesses and stores as places to buy things, schools where children are educated, parks where people can play or relax, and farms where food is grown. Younger students can draw pictures of land use while older students might write a few sentences or short paragraphs, or give a presentation on land use in their community.

Extensions: 30–45 minutes each

- If space is limited in your classroom, create a land puzzle instead of a paper-bag city. First, cut large sheets of butcher paper into seven or eight irregular pieces like a jigsaw puzzle. Then assign a different land use—such as homes, parks, mines, farms, and businesses—to each student to draw on the pieces. By fitting the pieces together, students can demonstrate how people work together to create good land use. Students could also discuss how some land uses are incompatible when put side-by-side. For example, a smelly garbage dump may be incompatible next to a park.

- Students can dramatize their city by role-playing mayors, police officers, firefighters, teachers, doctors, and farmers. Designate a city hall, police station, fire station, school, hospital, and farm in the paper-bag city. Consider different emergency scenarios such as a fire or a storm and ask how students would help one another.

- Tape strips of poster paper together and create a wall mural of a town or city. Students should remember to show gardens, parks, and farms.

National Science Teachers Association

Plant a Plant

Lesson Description

Students conduct experiments with growing conditions and raise their own plants. Students learn what plants need in order to grow, and learn what plants provide for humans.

Teacher Background

Sun, soil, water, and air—and *nutrients* provided by these elements—are essential for the healthy growth of plants. Plants use light from the Sun to conduct *photosynthesis* to survive. From air and water plants absorb hydrogen, oxygen, and carbon. From soil, lime, and fertilizers they obtain other essential nutrients, including nitrogen, phosphorous, and potassium. Students will perform classroom experiments to demonstrate the importance of soil and water in providing nutrients to growing plants.

Draw a parallel between plants and people: point out to your students that just as humans need certain nutrients for healthy growth, so do plants. (Here, you may wish to distinguish between *food* and nutrients.) Each nutrient helps plants in special ways. Some nutrients increase seed production and leaf, stem, flower, and root growth. Other nutrients hasten plant maturity and protect plants against extreme temperatures and disease.

Subjects

Art, Language Arts, Mathematics, Science

Time

Prep: 30 minutes
Activities: 2–3 hours
(not including Extensions)

SCI*LINKS*
THE WORLD'S A CLICK AWAY

Topic: plant growth
Go to: *www.scilinks.org*
Code: DIG06

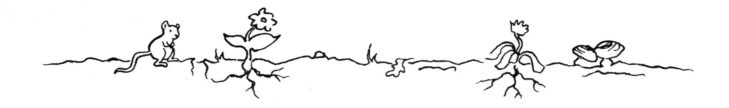

Student Objectives

Students will be able to:

• describe how most plants grow in soil; and

• explain that plants also need sunlight, water, and nutrients for growth.

Materials

For the Class

• Four potted and sprouting bean plants
• Four dishes or small trays
• Large tray
• Table or shelf that receives sunlight, or a grow light
• Watering can
• Water

For Each Student Group

• Cup of silty soil used in Lesson 1 (or see page x)
• Clear plastic cup or small plastic jar
• Permanent marker
• Two lima beans, mung beans, or peanuts
• Small plastic sandwich bag
• Rubber band
• Spray bottle
• Water
• Ruler
• Drawing paper
• Graph paper
• Pencil
• Two copies of Student Handout 6A

Soil gives plants the nutrients they need to grow. Plants in turn are food for animals, they produce oxygen for animals to breathe, and they provide materials for humans' food, shelter, clothing, medicines, fuel, and chemicals.

Learning Cycle

Perception: 30 minutes–1 hour

Prep About one week before you plan to teach the lesson, start growing four plants from lima beans, mung beans, or peanuts. For local sources of rapidly germinating seeds, contact your local Natural Resources Conservation Service office (see Appendix B). Use silty soil or soil with an equal percentage of silt, sand, and clay to grow the plants—do not plant the seeds in straight sand or clay.

Number the pots 1, 2, 3, and 4. Make sure there are holes in the pots for water drainage, and water the plants when needed. Place each pot in a dish or tray to catch water that may drain out of the pots. Set the pots in sunlight or under a grow light.

1 Introduce this lesson by showing students four potted bean plants.

2 Ask students to name the things plants need to grow. (Answer: air, water, and light.) Emphasize that green plants make their own food using the energy from light, and that soil provides the other necessary ingredients for growth. Nutrients and minerals dissolved in water enter plants through the roots in the soil.

National Science Teachers Association

3 Discuss how to conduct an experiment to prove that light and water are essential for plant growth. Ask students to predict what will happen to the four plants if one plant is put in a dark closet without light and water, one in a dark closet without light but with water, one in light without water, and one in light with water.

4 Distribute copies of Student Handout 6A and ask students to circle the correct words for each test environment. Then have students draw their predictions for each plant. You might ask older students to label their drawings and write one or two sentences about the test environments. (Students will need clean copies of this handout for the post-activity test in the Evaluation section.)

5 After discussing predictions, place the four plants accordingly and explain that the class will observe these plants daily for a week.

6 Water the "wet" plants as necessary over the course of the week.

Exploration: 30 minutes

Prep Before you conduct this activity, soak the beans in water overnight to speed germination. Poke small holes in the bottom of the plastic jars or cups for drainage.

Each student work area should have a clear jar or cup three-quarters full of soil, two beans, a permanent marker, a spray bottle with water, a sandwich bag, and a rubber band.

1 Demonstrate how to correctly plant a bean in a clear plastic jar or cup: plant the bean near the

**Figure 6.1.
Planting a bean.**

side of the jar so the growing roots will be visible. The bean should be covered with at least two centimeters of soil, and there should be space between the soil and the top of the cup (see Figure 6.1). Soil should be moist but not wet.

2 Have groups plant two beans in plastic jars and water the soil.

3 Ask each student group to write the group members' names on the cup, then carefully cover the cup with the plastic bag and secure the bag with the rubber band.

4 Place all the cups on a tray, and place the tray in a window or under a grow light

5 Clean up work areas. Save unused soil for Lesson 10.

6 Students should check their plants daily and add water if necessary. For the first few days of plant growth, students should only mist the soil if it is dry. Remove the sandwich bag as soon as the first shoot appears (approximately three to four days after planting).

Application: 30 minutes–1 hour

1 Throughout the week, students should observe their own plants. Students can measure and record daily plant growth on a data table that you create for the class beforehand, or on drawing or graph paper.

2 At the end of the week, help students create a simple graph of plant growth. (Younger students may be able to only compare two stages of growth and may need help to recognize the parts of a plant.)

National Science Teachers Association

3 If weather permits, take students outside to discuss other things that affect plant growth besides soil, water, and light (e.g., pollution, human interference, and air). If you conducted Lesson 1, review what happened when water was added to soil in a jar. (Answer: air bubbles rose, indicating that soil contains air. Plants need the carbon dioxide in air to conduct photosynthesis.)

4 Guide students to consider:

❑ how do the seasons affect plant growth?

❑ what happens to plants during storms and high winds?

❑ what happens to plants when there is too much or too little moisture?

❑ what happens to plants when there are too many insects and weeds?

❑ what happens to plants in soil that is nutrient-poor, rocky, sandy, or made of heavy clay?

5 Discuss farmers' and gardeners' work with plants and soil. What elements do farmers and gardeners try to control? (Answer: insects, nutrients, and weeds.) What elements can't be controlled? (Answer: rainfall, wind, and temperature.) How are nutrients added to the soil? (Answer: the farmer or gardener works fertilizer—along with air—into the soil. How does a farmer or gardener control weeds and insects? (Answer: cultivating plants, and using chemical and biological controls.) This discussion will help students understand what plants need to grow and how humans control the landscape.

Evaluation: 30 minutes

Students should be able to list several things that plants need in order to grow. Distribute new copies of Student Handout 6A and have students draw and label the four test plants after one week in the test environment. Students can look at plant growth over time by comparing the old and new handouts. Students can also write a sentence or a paragraph about the effects of the test environment on each plant.

Extensions: 30 minutes each

- Read the story "The Trees Speak" that follows this lesson, and color in Student Handout 6B.

- Read *The Giving Tree* by Shel Silverstein (see Appendix B). This short book tells the story of a tree that gives all that it can to a boy.

- Show the film or read the Dr. Seuss classic *The Lorax* (see Appendix B). This tale depicts economic greed and the abuse of trees, and describes how to help conserve natural resources.

- Students can take their bean plants or the potted plants home to plant in their yards and gardens, or keep in a pot inside the house.

- Plant flowers in a cup or container and take them home, or to patients in a nursing home or hospital.

- Write a play about growing plants.

- Have students pull old socks over their shoes and go for a "sock walk" in a nearby field or park to collect seeds that stick to their socks. Plant the seeds in containers to see what kinds of plants caught a ride.

National Science Teachers Association

Name: _____

Test Conditions: Dark / Light
Wet / Dry

Test Conditions: Dark / Light
Wet / Dry

Test Conditions: Dark / Light
Wet / Dry

Test Conditions: Dark / Light
Wet / Dry

National Science Teachers Association

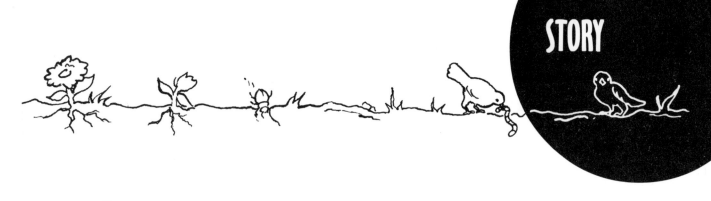

The Trees Speak

Teacher Tips

Trees do much more than provide products for human consumption. Thanks to their intricate root systems and leafy branches that catch rain drops and buffer the wind, trees are Earth's great protectors. They help hold the soil in place, thereby conserving soil that otherwise would be washed or blown away. There must be soil to produce most plants, and there must be plants to prevent soil erosion.

"The Trees Speak" teaches students to care for these valuable resources. This story will take your class on a magical trip to the woods where Peter, the hero, learns about all the ways that trees help humans. Peter and your students will discover that each tree is different, that trees help humans, and that replanting and recycling trees protects a natural resource.

You can conduct many activities to reinforce concepts in "The Trees Speak." After reading this story direct student conversation to the many ways people use trees. Ask your students to identify classroom products made from trees, such as paper, pencils, desks, and chairs.

Students can brainstorm how trees help humans. You can also discuss how trees play an important role in the environment by providing habitat for animals, cleaning the air, holding soil in place, and playing a part in the hydrologic cycle. Remember to emphasize that most plants (except air plants, aquatic plants, and marine plants) depend on soil as the home for their roots—roots hold soil in place, reduce erosion, and absorb dissolved minerals and nutrients that feed the plant. If weather permits, take your class outside and sit under a tree for a discussion.

Have your students act out what it is like to be a tree—e.g., bending in the breeze, losing leaves in the fall, and growing from a seed to maturity. A narrator can explain how trees obtain nutrients while students act out roots, water, soil, and sunlight.

Students can color in Student Handout 6B, draw a picture of a world without trees, or write a poem about a tree.

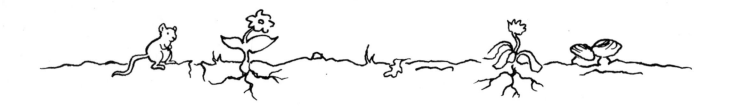

The Trees Speak

Darren leaned his head against the back of the porch and counted to 50. "Ready or not, here I come," he announced.

Peter had already run off into the woods away from his friends Sally, Julio, and Darren, and was well hidden. "Darren will never find me here," thought Peter as he sat down beneath a large oak tree.

Peter leaned his back against the tree and closed his eyes. He liked "Hide and Seek" but they had been playing for over an hour. Peter was getting tired.

"I'll just rest a minute," Peter said to himself, as his head fell forward and his chin dropped to his chest. Peter didn't know how long he stayed in this position. It must have only been a few moments because he almost immediately heard voices.

"Look at him," said the first voice, "sound asleep to all the cares of the world."

"He's rather sweet when he's asleep," said another voice with a much higher tone.

"That's the answer," laughed a third voice in a deep echoing tone. "Let's find something that will put them all to sleep for the next 200 years."

Several others must have found this very amusing because the woods were suddenly filled with the sounds of laughter.

Peter sat up and rubbed his eyes. He looked to the left, then to the right. He looked all around, but there was no one there.

"Look," said the first voice, "he's awake. You know it's against the rules to awaken humans."

National Science Teachers Association

Peter sat up straight and pressed his back against the tree trunk. "Darren, is that you? Julio? Sally?" he said.

"Of course not," said the deep voice, "I am speaking to you. And I wish you wouldn't lean so heavily against me. You're giving me trunk trauma."

Peter jumped to his feet. "I didn't know trees could talk."

"Listen to this kid," said a bushy cottonwood tree. "This child of asphalt probably doesn't know that fish can swim and birds can fly either."

"Ah, well," sighed the willow tree with the soft voice, "what can you expect from someone who watches television only with his eyes."

"My sister is now a television cabinet and encloses the TV with her whole body," explained the walnut tree, proudly. "She's also an end table, a dining room chair, and part of a bookcase."

"I'm afraid for my family," said the giant redwood, looking down on all the trees. "There used to be forty different types of me, now there are only three."

Peter was shocked. "What happened to the others?"

"They became your floors, bookshelves, and tables, little human. They became your toilet paper and notepads, toothpicks and newspapers. You name it," answered the redwood, gruffly.

"I don't want to brag," said the dogwood, "but think of all the things we do for you."

"Our leaves blanket the Earth and warm it from the chill winter winds," said the buckeye.

"Our greenery helps make rain," boasted the ash.

The pine shook its needles. "Our roots dig deep in the ground and keep the soil from shifting."

"We breathe in your carbon dioxide and give you oxygen in return," rasped the old hazelnut.

"Our wood becomes your homes and furniture," said the oak.

"And we feed you," cried the apple, cherry, peach, orange, and pear trees in unison.

"Our branches are the homes of small mammals and birds, and some of us give you magnificent beauty," said the lovely aspen, her golden leaves quivering in the sunlight.

"And," roared the crab apple, "you cut us down and grind us into pulp so you can have an extra magazine in your mailbox."

The large elm bent down to Peter. "Excuse my angry bothers and sisters," he said gently, "but they're right. If you overuse us and don't replant us or recycle our products, the Earth will become a very sad place."

"Try to think of the world without trees," sighed the tiny Japanese plum, her little purple leaves shaking in fear at the idea. "Why, the ground would start shifting this way and that, because our roots wouldn't be there to hold the soil in place."

The pine shook his needles. "And without us holding it in place, the rain would wash away the soil you use to grow your food," he added.

"The Sun could scorch the land dry if there were no trees to give shade," said a mighty maple.

Peter was very upset by what he heard. He sat down and tried to imagine such a world.

"Peter," someone shouted. "Where are you, Peter?"

Darren, Sally, and Julio came running through the forest. "Oh, there you are," said Sally. "We've been looking for you everywhere."

Peter jumped up excitedly and pointed at the oak. "Trees speak," he announced. "I heard them. I've been talking to them."

His friends thought that was very funny, and they laughed and laughed.

"It's true," said Peter, stamping his foot in anger. "Trees do speak. Just listen and you'll hear them too."

The children were very quiet. They closed their eyes and listened, but except for a soft fluttering of leaves, the trees were silent.

"That's OK," said Peter, shrugging his shoulders. "Maybe the trees want me to speak for them." He led his friends out of the forest. "You see," he said, "trees do a lot of things for us. Just try to imagine what this world would look like if there weren't any trees."

Peter talked all the way home.

Name: _____

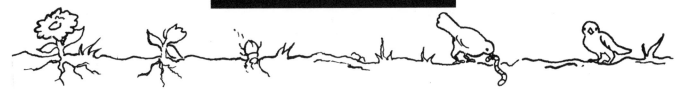

What Lives in Soil?

Overview

This section introduces students to the animals that depend on soil. In addition to studying the animals students can see, introduce them to Earth's vast invisible life—the billions of microscopic organisms, such as bacteria, protozoa, fungi, and algae, which inhabit and enrich the soil. In a single spoonful of fertile soil there may be seven billion bacteria, five million protozoa, a million fungi, and one hundred thousand algae. That's more than the total number of people on Earth!

These tiny microorganisms, and larger visible animals, contribute to a whirl of interrelated activity that builds, enriches, and restructures the soil. Burrowing worms and mammals, such as moles, badgers, and prairie dogs, help keep the soil loose and crumbly and allow water to penetrate and circulate in the soil.

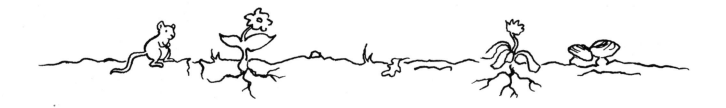

Loose soil, air, and water created by burrowing animals make it easier for plant roots to take hold and grow. Microscopic organisms decompose dead insects, larger animal matter, leaves, twigs, grass, and fallen trees. The rich organic humus that results enriches the soil for living organisms. Each organism is bound to the others in the intricate web of food chains, which ultimately depend on the soil.

Animal Apartments

Lesson Description

By dividing the soil environments into a model of an apartment building—with inhabitants on the basement, ground level, and upper levels—students learn what animals make homes in each soil habitat.

Teacher Background

This lesson demonstrates the amazing variety of animals that live in, on, and above the soil. Students examine small, individually assigned outdoor areas for terrestrial animal life, and report their findings. Class discussions about the habitat and food of these creatures will show that in one way or another, all living things depend on the soil for nutrition and survival.

Living things are found almost everywhere in the world. However, where they live depends on what *resources* they can obtain from the *environment*. All organisms—including humans—need air, water, and food to survive. Environments provide different kinds of food and climate, and therefore support different forms of life.

The differences between environments can be as small as the difference between the soil surface and the

Subjects

Art, Language Arts, Science

Time

Prep: 30 minutes
Activities: 1 ¾–2 hours
(not including Extensions)

SCILINKS.
THE WORLD'S A CLICK AWAY

Topic: microorganisms in soil
Go to: *www.scilinks.org*
Code: DIG07

Student Objectives

Students will be able to:

- gather data on the animals that live on the soil;
- describe the animals that live in, on, and above the soil; and
- define animal habitats.

space below ground. The animals in Figure 7.1 may be found in the various soil environments—in, on, or above the soil—near your school. Many of these animals must inhabit more than one soil environment to obtain the resources necessary for survival. Not all of the animals in Figure 7.1 will be found in all parts of the country, or at all times of the year. You will have to adapt this table for your own region.

Figure 7.1. Animals found in soil environments.

Found in Soil	Found on Soil	Found Above Soil
earthworm	snail	butterfly
ant	slug	moth
beetle	spider	dragonfly
insect larvae	millipede	bee
chipmunk	grasshopper	mosquito
mole	ladybug	tree frog
badger	snake	chickadee
prairie dog	turtle	nuthatch
groundhog	frog	hawk
gopher	lizard	woodpecker
rabbit	pigeon	jay
fox	robin	squirrel
mouse	sparrow	bat

National Science Teachers Association

Learning Cycle

Perception: 30 minutes

Prep On a large piece of poster paper, sketch an apartment building with at least three levels: basement, ground level, and second level (see Figure 7.2, which is a small version of one of the Student Handouts).

1 Referring to your sketch, discuss the floors in an apartment building (basement, ground level, second level). Ask students how an animal's home is like a person's home, and help students understand that animals make their homes in a similar way in, on, and above soil. Explain that today's activity will examine some organisms that live in the basement levels of soil—below ground.

2 Ask students what animals live below ground. Discuss habitats and environments and why some animals burrow underground for safe nesting. Discuss how decomposing plant and animal life in the soil provide nutrients for plants. (This sets the stage for Lessons 8 and 9 about food chains and decomposers.)

3 Write down students' suggestions on index cards—one animal per card—and tack the cards next to the basement level on your poster. Review the relationship between basement residents and organisms in the soil.

Materials

For the Class
- Poster paper
- Markers
- Index cards
- Pushpins
- Instant or digital camera (optional)

For Each Student Group
- Yarn (approximately 0.9 meters)
- Small plastic hand magnifier (approximately 5x magnification)
- Drawing paper
- Pencil
- Student Handouts 7A and 7B

Figure 7.2. The environments in an apartment building.

Exploration: 30 minutes

Prep Make copies of Student Handout 7A, which has drawings of animals commonly found on the soil in many parts of the country. (The drawings in the handout are not to scale.) Do library or Internet research to find common animals native to your area, and photocopy drawings of those animals. You might also find pictures of animals in biology textbooks, encyclopedias, or through newsletters and magazines of local wildlife conservation groups.

If your students have had no prior experience with studying animals you may wish to spend extra time discussing habitats and animals before this activity.

1 Distribute Student Handout 7A and pictures of animals native to your region. Discuss the animals that students will observe outside on the "ground level" on the soil. Point out the tiny size of many organisms and elicit suggestions on how to become more aware of these living things.

2 Take the class outside to a grassy area. Distribute hand magnifiers, paper, pencils, and yarn.

3 Demonstrate how to lay out a yarn lasso in the grass and use a magnifier to examine all the animals within the study circle, then have students do the same. The circle of yarn enhances close-up study and prevents running around or clustering around another student's discovery area. Caution students to observe but not touch any living animals or plants.

4 Have students draw what they found in their study circles. This will help students recall their discoveries as they share with the class. If you have access to an instant or digital camera, take a photograph of each student's study circle. The photos can help students see size differences and make comparisons.

National Science Teachers Association

5 Gather students and sit quietly to observe larger animals on the soil, such as mice, squirrels, chipmunks, frogs, snakes, pigeons, robins, and sparrows. If you don't see any animals, ask students to name animals they've seen before on the soil surface.

6 In the classroom, write down all students' discoveries on index cards and pin the cards next to the ground level on the animal apartment poster.

Application: 30 minutes

1 Review what students discovered about basement-level and ground-level dwellers in the animal apartment.

2 Ask students to name animals that live above the soil.

3 Write down all students' discoveries on index cards and pin the cards next to the upper level of the animal apartment illustration.

4 Ask if any of these upper-level organisms ever move down to the ground level. For example, why would a squirrel move from above the soil down to the soil surface? (Answer: to get food or nesting material.) Why would an owl swoop to the ground even though it roosts in a tree? (Answer: to catch a mouse.) Why would a butterfly land on a flower? (Answer: to get nectar or lay eggs.)

5 Help students understand that many animals must cross habitat levels to get resources such as food, just as a person living on the second floor of an apartment building has to go the ground floor to buy food from the market. Some of the animals students found on the soil may actually live

primarily below the soil (e.g., earthworms) or above the soil (e.g., birds).

Evaluation: 15–30 minutes

Distribute Student Handout 7B. Students should draw two animals living at each level and label the drawing. You may wish to provide younger students with a list of words or names to choose from. Older students can draw more examples, or write sentences describing the habitats.

Extensions: 30 minutes each

- Read a story about animal habitats from the resources list in Appendix B.

- Set up an ant farm so students can investigate ant life in the soil.

- Have students write and illustrate a book on animal apartments to share with other classes or visitors (possibly during a special event focusing on the environment or on soil and water conservation).

- Find a place where students can dig into soil (or find a place where a deep cut has been made into soil) to see what the basement level—beneath the soil surface—looks like.

Name: _____

snail

earthworm

grasshopper

spider

ladybug

beetle

butterfly

frog

snake

turtle

lizard

chipmunk

sparrow

robin

gopher

squirrel

mouse

Name:

Living Links

Lesson Description

Students learn about food chains and how humans fit into such chains.

Teacher Background

This lesson introduces *food chains* and explains soil's critical role in supporting life on Earth. One way to think about the organisms in an environment is by using a food chain. Some animals eat plants for *food*, while other animals eat animals that eat the plants. Plants are therefore an integral part of any food chain, and plants in turn depend on the soil to provide *resources* for survival. The soil food chain is the community of organisms living all or part of their lives in the soil.

Living things continue to be part of the food chain even after they die. Some animals are *scavengers* or *decomposers* and depend on dead plant and animal matter for food. (In Lesson 9, your students will learn more about one of the most familiar decomposers in soil, earthworms.) *Organic material* enriches soil and provides *nutrients* for soil-dwelling animals and plants.

Subjects

Art, Language Arts, Science

Time

Prep: 15 minutes
Activities: 2–2 ½ hours
(not including Extensions)

SCI LINKS
THE WORLD'S A CLICK AWAY

Topic: food chains
Go to: *www.scilinks.org*
Code: DIG08

Student Objectives

Students will be able to:

- describe a simple food chain;
- construct a food-chain mobile based on their understanding of food chains; and
- understand that all living things need soil to produce their food, either directly or indirectly.

This lesson dispels young learners' misconception that our food originates in grocery stores. For instance, consider the chain of events needed to make breakfast cereal: the cereal, milk, and fruit were purchased in the grocery store, but before that, those items came from plants, cows, and trees. The grains used in cereal, the fruit trees, and the cow's food were grown in soil.

Materials

For Each Student Group

- Index cards
- Cardboard
- Glue
- Single-hole puncher
- Crayons, colored pencils, or markers
- String
- Scissors
- Dowels or sticks
- Student Handouts 8A and 8B

Learning Cycle

Perception: 30 minutes

Prep On the board or on poster paper, sketch a simple food chain like the one in Figure 8.1. Draw arrows from predator to prey to show who eats whom in the chain. Use animals and plants native to your region.

1 Ask students to explain what the sketch represents. Guide students to understand that animals eat plants or other animals, and that a

Figure 8.1. A food chain.

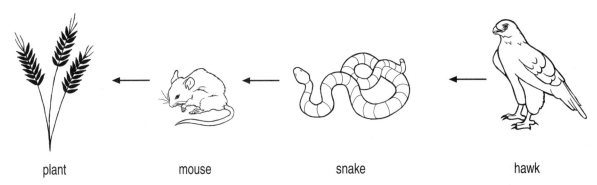

plant   mouse   snake   hawk

National Science Teachers Association

food chain connects animals and plants based on who eats whom.

2   Ask students how dead animals and plants can be part of the food chain. Discuss scavengers or decomposers that depend on dead plant and animal material for food.

3   Help students understand that plants are the basis for most food chains: almost all kinds of animal food can be traced back to plants. In the food chain in Figure 8.1, the snake doesn't eat plants but it eats the mice that eat plants. Discuss how the Sun, rain, and soil fit into the food chain by providing nutrients that plants need to grow. To emphasize this point, you might add a Sun at the top of the food chain sketch and a strip of soil at the bottom.

Exploration: 30 minutes

1   Design a new food chain based upon the animals discovered in Lesson 7, or have students suggest animals and plants to use. Keep the original food chain sketch on display while you write students' suggestions on the board. Encourage class participation, and keep answers general. The purpose of this lesson is to stress that living things depend on the soil, rather than to emphasize exactly what each animal eats.

2   Repeat this excercise using other examples of living things. Discuss the foods that students ate recently. For example: (1) Sun, water, and soil make grass grow; (2) a cow eats grass; (3) the cow produces milk; and (4) humans consume milk. Cows also provide nutrients for plants through their exhaled air (carbon dioxide) and other waste products that enrich soil.

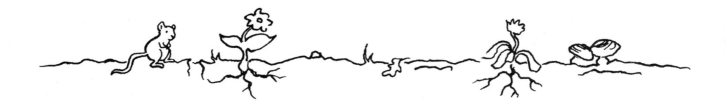

3 Encourage students to brainstorm how the parts of a recent meal were grown. Lead them through several examples of how each item arrived on their plate.

Figure 8.2. A food-chain mobile.

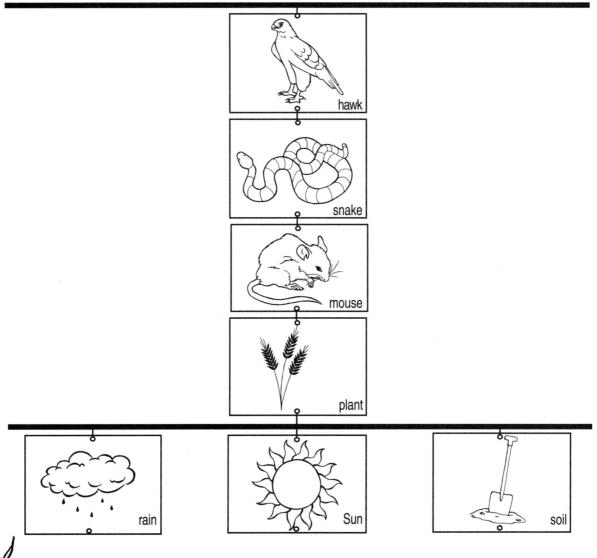

Application: 45 minutes–1 hour

1   Explain how to construct food-chain mobiles by putting one feature on each index card (see Figure 8.2). Older students can design their own mobiles, but give younger students a list of animals and plants to choose from to construct their mobiles. If mobiles are too complex for your class, have students draw animals and plants on one piece of paper and connect drawings with arrows. Older students might make more sophisticated mobiles with several levels, including decomposers and multiple food sources.

2   If your class is making mobiles, have students draw and label each feature of their food chains on a separate index card. In addition to animals and plants, students could include environmental features such as the Sun, soil, water, and grass as shown in Figure 8.2. When the drawings are complete, students glue the cards to cardboard.

3   Students put the cards in the proper order and then punch a hole in the top and bottom of each card.

4   Using the cards, string, and dowels or sticks, students can construct their mobiles.

Evaluation: 15–30 minutes

Students can label the illustrations on Student Handout 8A. You may wish to provide younger students with a vocabulary list or choice of words to use when labeling the drawing. Older students can write sentences describing each food chain event.

Extensions: 30 minutes each

• Read the story "Sara, Queen of Corn" that follows this lesson and color in Student Handout 8B.

• Play a game called "Are You Related?" in which students determine how an organism or feature is related to others through a food chain. For example, you might write "Sun," "grass," "cow," and "human" on different cards. Each student receives a card and role plays the plant, animal, or environmental feature named on the card. Split students into groups of four. One group at a time, students must explain to the class how they are linked to each other. This can be played over and over because many different chains can be formed from a large group of cards.

• Give each student one of the cards with an organism or environmental feature on it. Position the student with the "Sun" card in the center of the floor. Students position themselves in a circle around the Sun, then stretch a string between features that are related (for example, Sun—water—soil—grass—worm—robin, etc.). The result is a web of life, illustrating interrelationships between animals, plants, and environmental features. You can also use humans as one of the features to show that humans are part of nature's relationships.

Name: _____

1 —
2 —
3 —
4 —

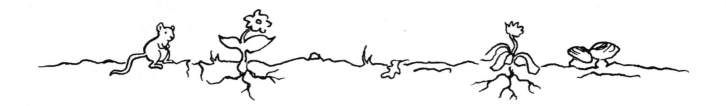

National Science Teachers Association

Sara, Queen of Corn

Teacher Tips

Corn, or maize, is a North American plant with many uses. "Sara, Queen of Corn" is the story of an imaginative child who lives on a farm and learns about corn. Reading this story will help your class learn about corn and other crops: where crops are grown, how crops are harvested, who harvests crops, and where crops go after the harvest.

After reading the story, ask students to describe daily life on a farm. If your school is urban or suburban and students don't have much knowledge of farms, consider bringing a farmer or gardener to your school to talk about his or her activities. Your class could develop a play around corn or another crop from your area. If so, be sure to include the natural resources such as soil, water, and sunlight and show the seasons, harvesting, and other aspects of agriculture.

You can develop several art activities around "Sara, Queen of Corn." Make a mural depicting the seasons, crops, harvesting, and other aspects of agriculture. Or, students can color the illustration on Student Handout 8B.

Bring in unhusked ears of corn and a picture of a corn plant to show students what corn looks like. You might even surprise your students with a treat of popcorn, just like at the end of the story!

Sara, Queen of Corn

Sara loved living on her family's corn farm. There was always something exciting going on. Sara was seven years old and had lived on the farm all her life.

"Raising corn is a very important thing to do," her parents told her. "There are a lot of hungry people out there who need our crop and the products made from it."

Sara pretended she was the Queen of Corn. Everyday she would walk among the stalks and tell them what to do.

"Make sure you eat your food to get vitamins," she would tell them. "Drink all your water," she would scold. "Grow straight and don't slouch."

Sara helped plant the corn, and she watched as it grew from little seedlings to plants that touched her knees. The plants grew quickly. After just a few months, they were as high as her waist, then her shoulders. The stalks grew so tall that even her father looked small when he stood beside them.

The ears of corn also started out very small, smaller than the tip of Sara's little finger. As the summer passed, the ears grew bigger and bigger. Slowly, the kernels turned from a ghostly white to a rich yellow, the color of the Sun that helped them live.

Although the corn was grown by Sara's family, there were many other animals that liked to eat the crop. Sara often saw mice running through the fields, nibbling on bits of corn that had fallen to the ground.

One summer night brought a terrible storm. The wind ran through the fields, and streaks of lightning lit up the sky. Sara held her hands over her ears to keep the sound of thunder from echoing in her head.

National Science Teachers Association

The family gathered nervously around the kitchen table, but it wasn't the thunder and lighting that frightened them.

"It's going to rain," said Sara's mother, "and there is nothing we can do about it."

"What's wrong with rain?" asked Sara. "I thought water was good for plants."

Sara's father put his arm around her and held her close. "There's nothing wrong with rain," he said. "But too much rain can be bad for the corn crop."

He looked down at Sara and laughed. "You know, Sara, you and the corn are very much alike. You both like to drink water, but you both hate to take a bath."

It rained that night, but it was only a light rain. The corn had a good drink and continued to grow.

Sara and the corn really were alike. Both were natives of North America. Both loved sunshine and warm summer evenings. Both needed careful, loving attention. And both the queen and her corn grew tall and strong during the summer months.

Growing corn, or any crop, is hard work. Yet there were always several days of farming that were pure fun for Sara. At harvest time, men and women came from all over the county with their large huffing and growling machines to pick corn and cut the stalks. It was like a big party.

Sara's family worked for days preparing for the great event of harvest. The workers had to be fed, so dozens of hams and chickens were baked. Sacks of potatoes and carrots were peeled, and the house was alive with the smells of cookies, candies, cakes, and pies. Sara helped with the cookies.

Sara awoke one morning to the sounds of dishes rattling and people laughing in the kitchen. She jumped out of bed and ran to the window. Today was the first day of the harvest, and she didn't want to miss a minute of it.

From her window she could see the tall stalks of corn swaying, their tassels moving back and forth. They looked like thousands of subjects waving yellow handkerchiefs at their queen. Sara quickly jumped into her shirt and jeans and ran down the stairs, out the door, and down to the field of corn. In the distance she could hear the sound of the great harvest machines crawling down the road toward their farm. It was difficult saying goodbye to such good friends as her stalks of corn, but the corn was needed somewhere else.

In a few days, most of the corn would be sent far away. Some would go directly to market to be eaten as fresh corn-on-the-cob, usually with lots of butter. Some would be made into corn oil at processing plants. Some corn would be made into corn bread or into hominy and grits. Others would end up in the canning factories where the kernels would be stripped from the cob and creamed for special lunches. The stalks and a few of the kernels would become food for animals. Some of it would even be shipped overseas to help feed people in other countries, and Sara felt good knowing her corn was going to help other people.

But there was one use for corn that Sara liked the most. The American Indians first invented this use, long before the people of other nations came to North America. For Sara, it came in handy on a warm summer's eve while sitting on the porch with friends, or when her parents took her to a movie.

Can you guess Sara's favorite use of corn?

It was to pop it.

National Science Teachers Association

Name: _____

National Science Teachers Association

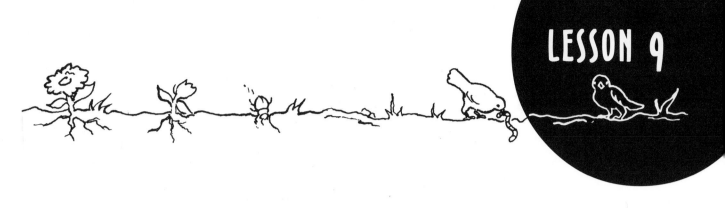

Watching Worms

Lesson Description

Students learn how worms eat, where they live, and how they help the environment. Students observe worms in soil and create art about worms.

Teacher Background

Earthworms benefit other animals, plants, and the soil because they are *nutrient* recyclers, or *decomposers*. Earthworms eat soil and digest the *organic matter* from the soil to get nutrients. The organic material came from plants and animals that lived in the soil. Earthworms spread their nutrient-rich excrement, or *castings*, on the soil surface and throughout the topsoil, creating new soil in which new plants can grow. The castings also provide food for smaller animals and microorganisms, which break down the organic material even further.

Earthworms live in deep, narrow tunnels, called *burrows*, in the soil. Earthworm burrows loosen the soil, allowing plant roots to grow down through the soil, and providing tunnels to the lower soil layers for smaller organisms. These burrows also *aerate* the soil, bringing oxygen down into the soil and permitting rainwater to flow through soil and carry nutrients to plant roots.

Subjects

Art, Language Arts, Science

Time

Prep: 30 minutes
Activities: 2 ½–3 hours (not including Extensions)

SCI*LINKS*
THE WORLD'S A CLICK AWAY

Topic: earthworms
Go to: *www.scilinks.org*
Code: DIG09

Student Objectives

Students will be able to:

- describe a worm's diet and habitat; and
- explain why worms are important for a healthy environment.

Materials

For the Class

- Tall, narrow glass or plastic jar with lid—e.g., tennis ball container
- Spoon
- Grass
- Dead leaves
- Soil (see page x)
- Water
- Bucket
- Two pieces of black construction paper
- Two rubberbands
- White crayon

For Each Student Group

- Newspaper
- Five Red Wiggler earthworms
- Tall, narrow glass or plastic jar with lid—e.g., tennis ball container
- Spoon
- Grass
- Dead leaves
- Soil (see page x)
- Two pieces of black construction paper

There are more than 7,000 species of earthworms. Earthworms live in temperate soils and in many tropical soils, and the different species can be found seasonally at all depths in the soil. Students may think these animals are "yucky," but they are necessary for healthy soil and a healthy environment.

Learning Cycle

Perception: 30 minutes

1. Review the soil layers discussed in Lesson 3. (If you did not conduct Lesson 3, information on soil layers is on page 17.)

2. Discuss worms that live in the upper layer.

3. Discuss a worm's diet and how it helps soil. Worms eat organic matter and recycle nutrients back to soil, to other animals, and to plants.

4. Discuss worms' habitat and that habitat's benefit to the soil. Worm burrows keep the soil loose, letting water and nutrients in and allowing plant roots to spread through the soil.

Exploration: 30 minutes–1 hour

Prep Obtain worms from a bait or vermicomposting (composting with worms) store. Before conducting the activity, keep the worms in damp soil and feed them composting scraps.

Cover student work areas with newspaper. Each work area should have a jar, a cup of damp (but not wet) soil, a spoon, composting scraps, rubber bands, one piece of black construction paper, and a white crayon. Poke a few small holes in the jar lids.

National Science Teachers Association

1 Students should fill the jar three-quarters full with loosely-packed soil. Students can then put bits of composting scraps into the jar on top of the soil.

2 Place five worms in each jar for the students. Have students cover the jars.

3 Demonstrate how to wrap the jar in black paper, leaving the top of the jar uncovered so that light comes in at the top and forces worms down into the soil. Students should secure the paper with two rubber bands, and use white crayons or colored pencils to write their names on the paper (see Figure 9.1).

4 Ask students to predict what will happen in their jars over time. You may wish to record student predictions.

5 For comparison with the students' jars, label two jars "Jar 1" and "Jar 2." Use dry soil in Jar 1 and damp soil in Jar 2. Add compost scraps but do not add any worms to these jars. Ask students to predict the difference over time between the three types of jars based on what you discussed in the Perception section.

6 Clean up work areas.

7 Your class can maintain these jars for several days, adding bits of composting scraps each day. If the soil is dry, add a little water to the jars. Students can remove the black paper from their jars to see how the worms have burrowed through the soil. Replace the paper after observation.

8 After several days, discuss the difference between student jars with worms and demonstration Jars 1 and 2 without worms. The soil in the worm-filled student jars should be dark and rich. The soil in

Materials Cont'd.

- Two rubber bands
- White crayon
- Eyedropper
- Water
- One piece of white construction paper
- Scissors
- Glue
- Writing paper
- Pencil
- Student Handout 9A

**Figure 9.1.
The worm jar.**

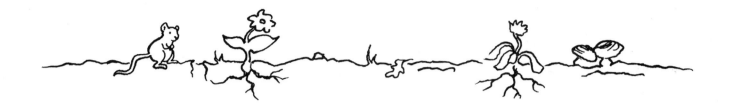

Jar 1 (unwatered, without worms) should look sandy and have dried-up compost; the soil in Jar 2 (watered, without worms), should look sandy and the compost moldy and not broken-down.

9 Distribute Student Handout 9A. Students should draw what happened in each jar. Students can label the drawings and note which compost scraps the worms ate. Provide younger students with a list of vocabulary words to choose from.

10 After your class has finished observing the jars, empty the jars into the bucket, than dispose of the waste material outside. Release the worms in a safe, protected place outside.

Application: 30 minutes

Prep Cover student work areas with newspaper. Each work area should have a piece of white construction paper, a piece of black paper, scissors, glue, and crayons, colored pencils, or markers.

1 Demonstrate how to cut out a large worm shape from a piece of white construction paper and glue it on to black paper to create a worm book (see Figure 9.2).

Figure 9.2. The worm book.

2 Students can write and illustrate short stories on the white space on their worm books. The story should describe life from a worm's point of view, describing what a worm eats, where it lives, and how it helps the soil.

National Science Teachers Association

Evaluation: 1 hour

Students can work alone or with a partner to design a poster that shows worms below the soil surface. Students should draw and label the three soil layers, along with plants and animals in the soil, including an earthworm. You might also ask your students to write on the poster useful things the worm does for the soil.

Devise a rubric that spells out your expectations for the poster and share it before you begin the project so that students know what is expected. Figure i.4 on page xvii shows an example of a rubric.

Extensions: 15–45 minutes

- Take the class outside to observe birds hunting worms. Watch a robin pause over the ground, detect a worm's movement, then strike the ground with its bill. Since an earthworm can't see or hear, it must feel the bird's vibrations on the ground to escape.

- After a rainstorm, look for worms on the wet ground outside. Help students understand that worms come to the soil surface during or after a rainstorm because water has filled their burrows and displaced the air.

- Take a field trip to a composting facility to see nutrient-recycling worms in action.

- If you teach older students, introduce the scientific method by having student groups design their own worm experiments. Students can write and then test a question, such as:

  ❐ What surfaces do worms like (e.g., light, dark, rough, smooth, shiny, etc.)?

  ❐ Do worms prefer light or dark environments?

  ❐ Do worms prefer dry or moist environments?

Name: _____

My Jar   **Jar 1**   **Jar 2**

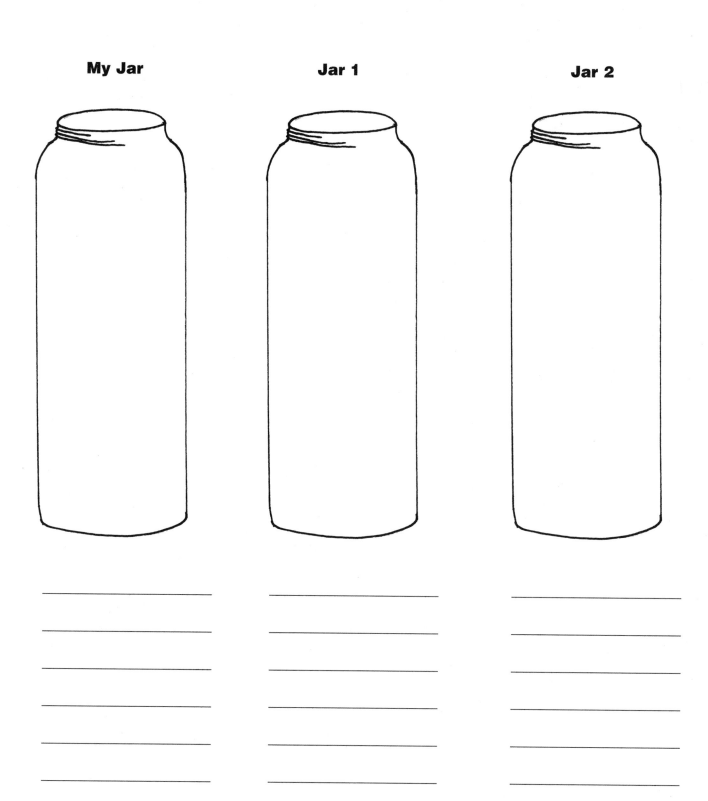

Where Is Our Soil Going?

Overview

Soil erosion can be caused by water or wind. Most soil erosion on American farmland is due to moving water. It is not easy to detect this type of soil erosion because runoff from heavy rains and melting snow and ice removes soil from Earth's surface in a very thin layer. Although undetectable at first, the natural process of erosion from water and wind can lead to significant soil losses from agricultural and urbanizing land. (Urbanizing land is that which is undergoing the transition from forest, grassland, and farms to roads and suburbs.) When soil is picked up by water or wind, the field from which it is moved becomes less fertile. Soil erosion increases the cost of farming because fertilizer need is greater, crop yields are lower, and farm profits diminish on eroded soil; this in turn increases the price of food.

Where does eroded soil go? Wind picks up soil and carries it in the air, while water moves soil into waterways. Those moving soil particles become sediment that lines roadsides and clogs our streams, lakes, and rivers. Soil sediment is the greatest water pollutant in the United States.

One particularly drastic example of soil erosion is the Dust Bowl drought of the 1930s. This drought and resulting soil erosion devastated farmland, destroyed the fertility of millions of acres, and filled the air with dust.

Droughts and wind will return to the Plains States, but the Dust Bowl storms are not likely to recur because many farmers, ranchers, and conservationists have learned how to minimize erosion. They have developed practices that conserve soil and water and protect the land's long-term productivity. Most importantly, they know the difference between soils that can be plowed without serious danger of erosion and those soils best left alone, and have restored native grasses to maintain soil fertility and water quality for future generations.

Going, Going, Gone?

Lesson Description

Students simulate rain and wind to observe the effects of water erosion and wind erosion on soil.

Teacher Background

Erosion is the loosening and movement of the solid material on the land surface by water *runoff*, wind, ice, and landslides. Wherever water flows or wind blows over unprotected soil, erosion is the result. Even on land protected by plants, some degree of natural erosion will occur. Soil continuously forms from *parent material* and *organic matter*, but soil erosion outpaces soil formation. The *sediment* that results from *water erosion* can cause water *pollution*. *Wind erosion* pollutes the air and reduces air quality.

When America's early settlers arrived in the area that is now the Plains States, they plowed the prairie grass to plant crops, not realizing that the native grasses held the soil in place. The exposed, plowed soil became vulnerable to the prairies' *droughts*, battering winds, and rain. Storms whipped up the dust, stripping the plowed earth of precious *fertile* soil, and destroying millions of agricultural acres. Thousands of farm and ranch families were forced to abandon their ruined land during the Dust Bowl droughts of the 1930s.

Subjects

Art, Language Arts, Mathematics, Science, Social Studies

Time

Prep: 30 minutes
Activities: 1 ½–1 ¾ hours (not including Extensions)

Topic: the Dust Bowl
Go to: *www.scilinks.org*
Code: DIG10

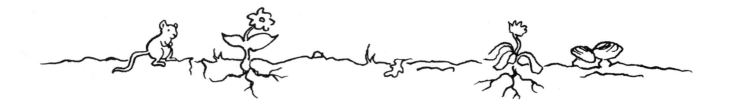

Student Objectives

Students will be able to:

- define erosion;
- demonstrate soil erosion using models of fields;
- explain where eroded soil goes and what its effects are; and
- determine how to protect land from soil erosion.

Materials

For the Class

- Color photos showing erosion
- Three small aluminum cake pans
- Dry soil (see page x)
- Dry soil with grass left on top
- Dry soil containing grass and plants roots and stems
- Three index cards
- Marker
- Three clear trash bags
- Hair dryer or mini fan
- Scissors
- Three measuring cups (at least 500 milliliters)
- Watering can
- Water
- Broom and dust pan
- Bucket

Tons of dry, powdery soils were carried thousands of miles by wind. Huge dust clouds were blown to the East Coast from as far west as Montana.

Soil erosion can never be stopped—it can only be controlled. In the lesson that follows, students learn about the effects of wind and water on bare soil, on soil covered by crop residue, and on soil protected by grass. Students analyze the results of erosion demonstrations to explore ways that *conservationists* treat the land to minimize soil erosion.

Learning Cycle

Perception: 15 minutes

Prep Find color photographs of erosion in Earth science and environmental science textbooks or encyclopedias, or cut out photographs from calendars or magazines such as *National Geographic, Audubon, Outside, Sierra Club, Journal of Soil and Water Conservation,* etc. If possible, find pictures showing the same land before and after soil erosion occurred.

1 Review the value of soil and discuss why soil is important to environmental health. (Answer: soil is the medium in which our food is grown and is the space where we build our towns and cities; soil provides habitat for animals, shelters plant roots, and gives animals and some plants critical nutrients for survival.)

2 Direct student's attention to the erosion photos. Help students understand how soil erodes from agricultural and urban areas through water and wind movement.

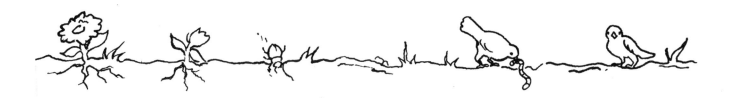

Exploration: 30 minutes

Prep Use scissors to notch a "v" in a short side of each cake pan (see Figure 10.1 on the next page). Prepare models of fields by filling one cake pan with bare soil (to represent a plowed field with no protection from vegetation), another pan with soil and loose plant material (to represent an agricultural field with crop residue left for protection), and the third pan with a patch of soil and firmly-rooted grass (to represent a pasture or meadow). Use a marker and index card to label pans "plowed field," "field with crop residue," and "meadow."

1 If your students have visited a farm or garden, discuss the kind of fields they've seen.

2 Ask students how the three cake pans are models for three different fields.

3 Explain that this experiment will simulate the effects of the weather and seasons on soil in those types of fields. Ask students to predict what will happen when "wind" blows across each "field." Have students give reasons for their predictions.

4 Distribute Student Handout 10A. Older students can fill in the first and second columns of the handout. Younger students can draw pictures of their predictions instead of writing sentences about what will happen, or make predictions for just one of the fields.

5 Select one student to hold a garbage bag open, and select another student to hold a pan and tilt it lengthwise over the garbage bag. Explain that tilting the pan models a sloped field.

6 Select one student to be the wind blower, or do it yourself. That student will hold the hair dryer

Materials Cont'd.

For Each Student Group
- Drawing paper
- Crayons, colored pencils, or markers
- Student Handouts 10A, 10B, and 10C

Figure 10.1. Simulating wind erosion on a field.

about 20 centimeters from the upper end of the pan, directing the blowing air down toward the garbage bag for 15–30 seconds (see Figure 10.1).

7 Repeat for each type of field.

8 Have the class gather around the garbage bags to see the results. The garbage bag under the "plowed field" should contain the most soil, while the bag under the "meadow" should contain the least soil. The bag under the "field with crop residue" should contain a medium amount of soil. Were student predictions accurate?

9 Discuss what these results mean. Help students understand that the grass in the meadow protects the soil from wind, while bare soil without any vegetation is exposed to wind and therefore the most vulnerable to soil erosion.

10 Students should complete Student Handout 10A by drawing pictures or writing sentences about what they observed.

11 Clean up spilled soil. Save pans for the Application section.

Application: 30 minutes

 Use the three pans from the Exploration section.

1 Challenge students to predict what "rain" will do to the three "fields."

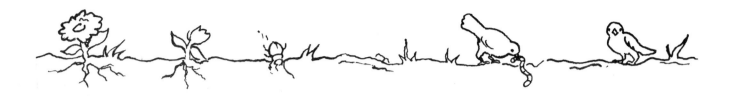

2 Distribute Student Handout 10B and ask students to draw what they predict will collect in each measuring cup. You may wish younger learners to make a prediction for just one pan, or to draw only what they see happen during the demonstration.

3 Select one student to tilt a pan—notch side down—as in the Exploration section. Select one student to hold the measuring cup just below the notch to catch the runoff.

4 Select a student or use the watering can yourself to "rain" onto the pan.

5 Repeat for each pan.

6 Allow the contents of the cups to settle. Help students read the marks on the cups to measure how much soil was lost from each pan. Measure the soil that has settled in the bottom of the cups, not the water or the organic matter that may float to the top.

7 Discuss the results of the demonstration. Guide students to understand that bare soil without any vegetation erodes the most, while soil with some vegetation remaining erodes the least, since plant roots hold soil in place.

8 Hold up the cup containing the most soil sediment and ask for ideas about where this sediment goes. (Answer: soil sediment may cover crops at the bottom of slopes, may be deposited in road ditches, may fill in lakes and swimming areas, may spoil fish, bird, and aquatic plant habitats, and may contaminate drinking water supply.)

9 Clean up the demonstration area. Collect all soil, water, and plant material in the bucket, then dispose of the waste material outside.

Evaluation: 15 minutes

To review the terms and concepts in this lesson, students can work on the word-find puzzle on Student Handout 10C. To make this more challenging for older students, don't provide a vocabulary list at the bottom of the page but instead provide clues about the word. Students fill in the blank with the correct word and then look for it in the word find. Answers to the puzzle are given in Figure 10.2.

Extensions: 15 minutes–1 hour each

- Take a conservation walk around your school yard to look for signs of erosion.

- Invite a conservationist to talk to the class about soil erosion. The conservationist may show students an area that has undergone corrective treatment, and explain the erosion treatment and its effects. If you need guidance, contact a soil conservationist at the Natural Resources Conservation Service (see Appendix B). In Lesson 11, students learn more about soil scientists and conservationists.

Figure 10.2. Answers to word-find puzzle.

```
P  F  E  R  T  I  L  E  N  E  F  I  E  L  D
D  A  M  A  G  E  Q  U  S  F  E  X  C  O  R
B  E  H  C  Z  N  O  D  O  Y  G  M  J  A  S
A  C  J  S  P  V  F  N  B  E  C  O  C  L  E
N  D  W  E  L  I  U  X  C  R  O  P  D  A  R
P  O  A  Y  C  R  G  L  H  A  N  T  C  F  O
A  R  D  B  S  O  M  I  Z  R  S  D  E  Q  S
S  P  H  J  X  N  K  O  N  L  E  S  P  J  I
T  N  L  C  G  M  Q  S  I  W  R  T  Y  G  O
U  E  Z  A  T  E  U  F  D  T  V  C  A  U  N
R  W  P  Y  N  N  A  G  D  N  A  L  N  W  W
E  S  I  H  O  T  N  O  B  I  T  X  A  K  C
A  L  T  N  A  U  C  U  J  Z  I  W  N  F  S
E  J  S  E  D  I  M  E  N  T  O  B  K  P  B
R  E  S  O  U  R  C  E  T  D  N  Q  Z  A  E
```

Name: _____

Field Description	Prediction	Observation	Were You Correct?
1			
2			
3			

Name: _____

Field 1

Field 2

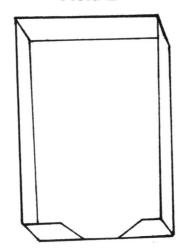

Field 3

_____   _____   _____

_____   _____   _____

_____   _____   _____

_____   _____   _____

_____   _____   _____

_____   _____   _____

Student Handout 10C

Name: _____

```
P  F  E  R  T  I  L  E  N  E  F  I  E  L  D
D  A  M  A  G  E  Q  U  S  F  E  X  C  O  R
B  E  H  C  Z  N  O  D  O  Y  G  M  J  A  S
A  C  J  S  P  V  F  N  B  E  C  O  C  L  E
N  D  W  E  L  I  U  X  C  R  O  P  D  A  R
P  O  A  Y  C  R  G  L  H  A  N  T  C  F  O
A  R  D  B  S  O  M  I  Z  R  S  D  E  Q  S
S  P  H  J  X  N  K  O  N  L  E  S  P  J  I
T  N  L  C  G  M  Q  S  I  W  R  T  Y  G  O
U  E  Z  A  T  E  U  F  D  T  V  C  A  U  N
R  W  P  Y  N  N  A  G  D  N  A  L  N  W  W
E  S  I  H  O  T  N  O  B  I  T  X  A  K  C
A  L  T  N  A  U  C  U  J  Z  I  W  N  F  S
E  J  S  E  D  I  M  E  N  T  O  B  K  P  B
R  E  S  O  U  R  C  E  T  D  N  Q  Z  A  E
```

Find the following words related to soil erosion:

conservation	fertile	resource
damage	field	runoff
crop	land	soil
environment	pasture	water
erosion	plant	wind

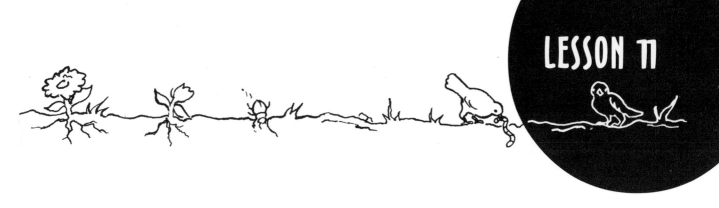

Soil Scientists

Lesson Description

Students learn how people in the community work to conserve natural resources.

Teacher Background

This lesson shows students what *soil scientists* and other *conservationists* do. This lesson may be more meaningful to students if you invite professional conservationists to the classroom. If possible, follow class discussions by a trip to an outdoor conservation site.

Opportunities in *resource conservation* are numerous because no single science can accomplish the great task of conserving our nation's *natural resources*. Moreover, such environmental issues are now global concerns. "Think globally, act locally" has become the common call of people from many nations who are working or volunteering to protect the Earth.

Conservationists enhance the productivity, safety, and beauty of our world. The many types of conservation professionals include agronomists, botanists, cartographers, ecologists, economists, engineers, foresters, geologists, hydrologists, plant materials specialists, public affairs officers, recreation managers, soil scientists, and wildlife biologists.

Subjects

Art, Language Arts, Science, Social Studies

Time

Prep: 30 minutes
Activities: 1 ¼–2 ¼ hours
(not including Extensions)

SCiLINKS
THE WORLD'S A CLICK AWAY

Topic: erosion
Go to: www.scilinks.org
Code: DIG11

Student Objectives

Students will be able to:

- identify what soil scientists do; and
- learn about resource conservation activities.

Materials

For the Class

- Magazine photos of conservationists, scientists, farmers, etc.
- Poster paper or bulletin board
- Tape, glue, or pushpins

For Each Student Group

- Drawing paper
- Crayons, colored pencils, or markers
- Writing paper
- Pencil

A visit from a soil scientist would be especially beneficial for the class to review the many kinds and uses of soil described in previous lessons (see Appendix B for information on contacting a soil scientist). Soil scientists inspect each soil horizon's slope, texture, color, structure, boundaries, thickness, and degree of erosion. To make accurate predictions of soil behavior, the soil scientist must learn all of the characteristics of a particular soil, because it is the unique combination of qualities that controls behavior. After collecting data, the soil scientist plots each sample site, identifies soil types, and outlines soils on aerial photos of the land. The finished work is a *soil survey* that helps farmers, ranchers, highway engineers, land use planners, homebuyers, and others decide how to use the land wisely.

Learning Cycle

Perception: 15 minutes

1 Discuss natural resources and resource conservation. Ask students for examples of natural resources and discuss why the resources are important.

2 Discuss the activities of conservationists, especially soil scientists.

Exploration: 15–30 minutes

Prep Cut out magazine photos of people working in conservation or on the land: conservationists, soil scientists, farmers, ranchers, wildlife biologists, botanists, engineers, ecologists, rangers, foresters, etc. Create a poster or bulletin board and label each picture.

1 Discuss the photos and the activities represented, and why they're important for the environment.

2 Ask students to think about these activities. Which activities would students personally like to do best?

3 Have students draw themselves involved in their favorite conservation activity. Ask students to add a caption describing what they're doing.

4 Create a conservation bulletin board with drawings from the class.

Application: 30 minutes–1 hour

Invite a farmer, soil scientist, or a conservationist to visit the classroom. Encourage students to prepare questions about the individual's activities. The resources list in Appendix B has more information on agencies and organizations involved in conservation.

Take younger learners on a field trip to a farm, nursery, or gardening center. You could also invite a gardener to come to class to give a demonstration on how to take care of a plant.

Evaluation: 15–30 minutes

Ask students to write thank-you notes to the visiting conservationist. In the letter, students should summarize some of the most important and interesting facts they learned from the visit and ask any other questions they might still have.

Extensions: 30 minutes–2 hours each

- Take a field trip to a conservation site. On a field trip to visit a soil scientist, for example, students can accompany the scientist to roadside cuts and pits to study soil.

- Encourage students to dress up as their favorite conservationists. Let students role play their part for the others to guess. This could be a special activity for parents' day.

- If appropriate for your class, students can research or write about a conservation activity.

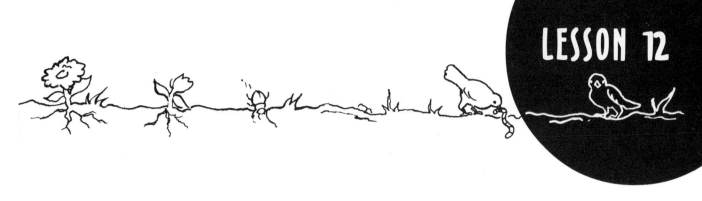

An Outdoor Learning Center

Lesson Description

With adults' help, students inventory the school site, develop plans, then create a garden. This will raise awareness among students, teachers, and parents about the natural environment and about using the school site for hands-on learning.

Teacher Background

This lesson provides students and teachers with an ongoing opportunity for hands-on environmental education and *resource conservation*.

A readily accessible resource for teaching is the school site. An outdoor learning center (OLC) on the school site offers educators and students an exciting place to observe nature's happenings through the seasons. Right outside the classroom, the school site offers many opportunities to publicize conservation in the neighborhood by improving students' knowledge about and concern for the natural world. By implementing an OLC on the school site, the teacher can maximize teachable moments relating to the environment and *natural resources*.

Subjects

Art, Language Arts, Mathematics, Science, Social Studies

Time

Prep: 2 hours minimum
Activities: 4 ½–10 hours
(not including Extensions)

If your school or neighborhood already has an OLC, skip ahead to the "OLC Activities" section below.

Enlisting Assistance

Begin the planning process as early as possible. It is important to secure the school administration's permission and obtain support from the school's maintenance staff for your project. If you teach young children, find a teacher of older grades who shares your interest in the project. This teacher's students and yours can become teammates or buddies for the school project. Remember to start small: the project has a greater chance to succeed if original goals are modest and leave opportunity for growth. As members of the school community see the success of this first step, they may provide support for an OLC for the entire school.

If your school site project has the space and has been well planned, it may be easily adaptable for additional outdoor learning activities in continuing school years. If this idea is approved by the administration, you may tell students, parents, and other teachers that this project will be the first step in establishing an OLC on the school site—a place to do hands-on activities, learn about the environment, and participate with actual resource conservation projects. Remember that you are dealing with natural as well as human influences. Be prepared to explain limitations of the OLC, such as temperature, moisture, insects, wind, limited space, or the wrong soil. Learning from this year's activities can help create a more successful OLC next year.

There are many ways to solicit the equipment needed to create and maintain your OLC. Team up with a high-school agricultural program and share supplies.

Apply for a grant with a local or national gardening or environmental education association. Ask a landscape firm, local business, or government agency to donate tools (see the resources list in Appendix B).

Creating an OLC Garden

A common and effective school site activity is establishing a garden, which is an environment that students can manipulate. Students' planning, planting, and caring lead to the excitement of harvesting the rewards of their efforts. You can establish a vegetable or flower garden almost anywhere: in a large or small space, on a flat area or on a slope, in the shade or in full sunlight, on the school roof, or on a narrow strip of land between a parking area and the school building. Whichever area is used, the OLC garden will provide a venue for short- and long-term environmental learning.

To create a garden, first analyze the site. Students should observe and record the site's physical and environmental characteristics. This class survey will provide a starting point and will show the changes that take place over time. Document modifications to the OLC garden to provide a compete record. After the site has been analyzed, discuss planting, maintaining, and harvesting a garden. The class can then decide on the type of flowers or vegetables to grow, design the garden layout, and plant.

Materials

Actual materials required for this activity will depend on the needs identified through the inventory and planning. A soil survey will determine the soil type of the school site and help you select the correct vegetation. Use native plant species whenever possible, since they tend to require less water, weeding, and fertilizer

than exotic species. Be sure that none of the plants are invasive, especially if your site is near any natural areas. You can obtain this information from local soil and water conservation agencies (see the resources list in Appendix B).

This activity offers a valuable opportunity to stress safety with your students. Emphasize the correct way to use and treat tools. For days when your class will be working outside, ask students to wear pants and shirts with long sleeves to avoid insect bites and irritating plants. Teach students what poisonous plants look like and how to avoid such plants. It will help to find out in advance which students have insect and plant allergies, and take necessary precautions.

OLC Activities

The activity, observation, and records of an OLC should be continual, and should demonstrate interrelationships between humans and the rest of the natural world. Activities should be inquiry-based and lead to the resolution of issues.

The following suggestions for OLC activities focus on *conservation*, *beautification*, and wildlife *habitat* improvement:

- Adopt a section of the OLC, a playground, or a nearby stream. Remove all trash and keep the area clean.

- Plant trees or shrubs that shelter the school site from the wind.

- Plant flowers, trees, or grasses to stop soil erosion.

- Invite birds to your area by adding birdhouses near shrubs or trees that provide protection from predators and by choosing plants that provide food and shelter (see Figure 12.1).

Figure 12.1. Plants that provide food for wildlife.

Trees	Shrubs	Flowers
Oak	Viburnum	Cosmos
Black Walnut	Blueberry	Impatiens
Crabapple	Dogwood	Marigold
Maple	Lilac	Zinnia
Pine	Sumac	Phlox
Spruce	Pfitzer Juniper	Trumpetvine
Desert Willow	Ocotillo	Desert Baileya
California Buckeye	Desert Hackberry	California Poppy

Some of these plants may not be appropriate for your region. Avoid using non-native plant species.

- Create a butterfly garden by using plants and flowers that attract butterflies.

- Order vegetation native to your area and plant a natural landscape.

- Plant grass and trees that are valuable for shade, nesting, beauty, and that vary in color, texture, and shape.

- Adopt a special tree and note seasonal changes, animals that live in the tree, and outstanding characteristics of the tree using photos, drawings, and writing.

- Identify rocks or boulders on the site. Investigate the types of materials used to build the school and compare materials to the rocks on the site.

- Examine a rotting log to observe fungi, moss, and insects.

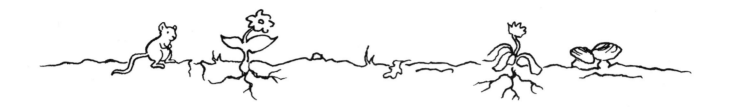

Student Objectives

Students will be able to:

- design and build an OLC garden;
- justify the importance of their school-site conservation activities; and
- explain some of the activities or events in the OLC garden.

Materials

For the Class

- Posterboard
- Marker
- Local soil survey
- Plastic transparencies
- Overhead markers
- Gardening tools (e.g., hoes, rakes, spades)
- Work gloves
- Plants, trees, and shrubs
- Hose
- Camera (optional)

For Each Student Group

- Diagram of the school site
- Pencil
- Writing paper
- Drawing paper

- Record temperature, wind, or precipitation over time, and then graph the data.

You might begin by having students classify the environmental events that take place on the school site on a regular and seasonal basis. Students can pass records to succeeding classes to build an environmental history of the site. Over time, students might chart differences in rainfall, snowfall, temperature, growth and death of plants, or erosion. Older students could research the history of the school site. By analyzing history and environmental events through tables, graphs, and written logs, students will become more aware of the school site environment.

Learning Cycle

Perception: 30 minutes–1 hour

1 Introduce students to the idea of an OLC.

2 Begin planning the project by brainstorming ideas for a garden. Let students lead by providing ideas and making notes on the board. What are the students' desires and concerns for the garden? Encourage students to discuss their ideas about planning and placement, and illustrate those ideas on the board and record them.

National Science Teachers Association

Exploration: 30 minutes–2 hours

 Sketch a simple diagram of the school site.

1 Take the class outside to map the school site. Distribute diagrams of the school site and have students record the physical characteristics of the site. For instance, you might ask students to map areas of bare soil, direct sunlight, vegetation, pavement, and buildings, and compare the slope of the ground in various places. To save time, you can assign each student group to map one characteristic of the site. Then transfer all the maps to clear transparencies and overlay the maps for an overview of the school site.

2 After students have created their maps, suggest to students how you will use this information to create a successful garden. Discuss how students' project ideas will work with the school site's available space. Adjust the plan, as ideas are accepted. This organizing session allows students to communicate, plan, and be responsible for the development of their own school site project.

3 To actively involve older students in the planning process, hold a contest to select the best plans for the garden. Divide students into groups of two or three and ask groups to draw up plans and materials lists of their ideas about what the garden should look like. A panel of teachers, administrators, maintenance staff, and older students choose the top three plans. The class then votes for its favorite plan out of the top three.

4 Using the class's suggestions, draw a plan of the garden on posterboard.

5 When the project has been finalized, type or print all relevant information and create the formal plan for the school site project.

Application: 3–6 hours

Prep Planning, organizing materials, getting permissions, and involving parents, students and school staff will take several hours. Be sure the adults don't take over the project; this should be a fun and exciting time for student discovery. Also, remember that there is no deadline—this project may never be finished. Ideally, the excitement generated by the school site project will encourage duplicate efforts in the school community and the community at large, starting with home gardens or other beautification activities.

1 Take the class outside and demonstrate the proper use of gardening tools.

2 Split students into small groups or pair students with older teammates/buddies.

3 Assign group roles and responsibilities. Some teams can begin planting while other groups sketch or list more ideas for the garden.

4 Every student should have the opportunity to do some type of gardening activity—raking, planting, etc. Such active participation gives students a sense of ownership for the program, and helps them develop a sense of belonging and personal satisfaction.

National Science Teachers Association

Evaluation: 30 minutes–1 hour

Evaluation should be an on-going process as the school site project is developed, and includes formal follow-up with students, parents, and other school staff. The students should be allowed to express suggestions for the next phase of the school site project.

Extensions: 30 minutes each, minimum

*   Read a story from the Resources section (Appendix B) about planting a garden.

*   As a class, discuss ideas for expanding the current school site project, develop a plan, and present it to the school administration.

*   If an OLC is not possible at your school, identify and label the vegetation currently growing on your school campus. Students could also observe wildlife on your school grounds or in a nearby park.

*   Students can grow vegetables or flowers in a garden, using stakes to identify each plant. Choose plants that will produce results before the school year ends.

*   If you grow produce in your school garden, try these ideas:

    ❏ Invite a parent to prepare some of the produce grown in the garden.

    ❏ Give produce to the school cafeteria to use in a meal for students.

❏ Donate produce to a homeless shelter or soup kitchen.

❏ Allow students to divide and take home any produce or flowers.

• Each student can pick a plant in the garden and measure and graph its growth over time. Students could also draw the plant in various stages of growth, or through the seasons.

• Establish a nature trail on or near your school site.

• Take a field trip to a farm or a garden center to see how "big" gardens are planted and cared for, or invite a farmer or garden center employee to class to share expertise, experience, and perhaps some plant materials or tools with students.

National Science Teachers Association

Glossary

aeration = the process by which soil is supplied with air, such as when a farmer or gardener tills the soil

beautify/beautification = to make something (such as land) more attractive

bedrock = solid rock that underlies the soil; also called *parent material*

burrow = a deep, narrow tunnel in soil made by worms and other animals

castings = nutrient-rich worm excrement

clay = the smallest particle of soil, less than 0.002 millimeters in diameter

clayey soil = very fine soil; lumpy if wet but hard if dry

conserve/conservation = the wise use of natural resources to prevent damage, pollution, and waste, therefore extending the life of the resources for use by future generations. Scientists who work to protect natural resources are called *conservationists.*

cross section = a piece of something cut open so that the inside is visible

decomposition = biological and chemical breakdown of nutrients from dead plants and animals, including bacteria, fungi, and other microorganisms. Worms are *decomposers.*

desert = arid region with sparse vegetation and less than 25 centimeters of precipitation per year.

dirt = soil out of place in the human world; a pejorative use of "soil."

drought = a period of dryness that causes extensive damage to crops or prevents their successful growth

environment = the interaction of physical, chemical, and biotic factors (such as climate, soil, space, and living things) that affect an organism's ability to survive

erosion = loosening and movement of the solid material on the land surface by water *runoff*, wind, moving ice, and landslides. Erosion can also result from humans disturbing the soil.

fertile = capable of sustaining abundant plant growth, rich in nutrients

food = a substance that nourishes a living organism

food chain = a series of plant or animal species in a community, each of which is related to the next as a source of food; also called a food web.

forest = area covered with trees and woody plants

habitat = a place where plants and animals live, grow, and reproduce

humus = rich *organic matter* that results from the disintegration of dead animals, leaves, twigs, and fallen trees in the soil

landform = a natural feature of the land surface; characteristic of a *habitat.*

land use = the varied ways that public space is used, such as for residential property, businesses, government buildings, parks, or recreation areas

model = a simulation of a real-world scenario

mountain = area with high elevation, rock material, and steep slopes; a type of *landform*

natural resource = material found in nature and used by humans, such as trees, water, and oil

nutrient = raw material that provides food for organisms' (including humans') growth

organic matter = decomposed plant and animal material, found in and on soil, that provides nutrients for living organisms

parent material = the unconsolidated and more or less chemically weathered mineral or organic matter from which soil develops

photosynthesis = the formation of carbohydrates from carbon dioxide and water in the chlorophyll-containing tissues of plants exposed to light

planning commission = a group of people in a community who are responsible for planning how the land in the area will be used

pollution = a condition, caused by substances in Earth's air, water, and soil, that reduces the quality of the *environment* for life

pore spaces = small spaces between soil grains that are filled with air and water

prairie = rolling or level grasslands with few trees and medium rainfall

resource = something that an organism must obtain from its *environment* to survive. Resources for animals include food, air, water, and shelter.

resource conservation = the preservation and protection of Earth's *natural resources*

runoff = water from precipitation that is not absorbed but flows over the land, carrying *sediment* and other materials to streams, lakes, and other bodies of water

National Science Teachers Association

sand = the largest particle of soil, between 0.05 and 2.00 millimeters in diameter

sandy soil = soil with sandy particles (larger and grittier than *silty soil*)

scavenger = an animal that feeds on dead animal matter or refuse

sediment = earth material carried by water from eroding areas of the land. Sediment can clog rivers and streams, destroy wildlife habitat, and pollute water supplies.

silt = the medium-sized particle of soil, between 0.002 and 0.05 millimeters in diameter

silty soil = soil with a high content of silt-sized particles; generally darker and looser than *sandy* or *clayey soil*.

soil = the collective term for the natural bodies of earthy materials that cover much of the Earth's surface; a complex combination of mineral and organic materials

soil scientist = someone who studies the types and properties of soil

soil survey = a map of the soil types in a particular region; a science-based inventory of the distribution and properties of soils and factors affecting the soil environment. Soil surveys include predictions of soil behavior related to selected *land uses* in an urban, agricultural, or natural environment, and the impact of land uses on these environments.

texture = the characteristic proportion of *sand*, *silt*, and *clay* in a particular soil

water erosion = the detachment and movement of soil by water

weathering = the breakdown of rocks and sediment at or near the Earth's surface due to biological, chemical, or physical actions

wetlands = a transitional area between water and land that is saturated long enough to support very moist soils and plants that grow in water

wind erosion = the detachment and movement of soil by wind

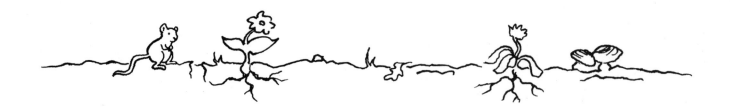

Resources

Suggested References

Children's Literature

 Bial, R. 2000. *A Handful of Dirt*. New York: Walker & Company.

Cole, J. 1986. *The Magic School Bus Inside the Earth*. New York: Scholastic, Inc.

McNulty, F. 1990. *How To Dig a Hole to the Other Side of the World*. New York: HarperCollins Children's Books.

Murray, P. 1995. *Dirt, Wonderful Dirt!* Chanhassen, Minnesota: This Child's World, Inc.

Nielson, S. 1993. *I Love Dirt*. Edina, Minnesota: ABDO Publishing Company.

Webb, A. 1986. *Soil (Talk About Series)*. New York: Macmillan/McGraw-Hill School Publishing Co.

 Baines, C. 1990. *The Old Boot—An Ecology Story Book*. Brooklyn, New York: Crocodile Books, Intenik Publishing Group, Inc.

Baylor, B. 1981. *The Desert Is Theirs*. New York: Simon & Schuster Children's.

———.1993. *Desert Voices*. New York: Simon & Schuster Children's.

Beskow, E. 1991. *Children of the Forest*. Edinburgh, Scotland: Floris Books.

Cole, J. 1995. *The Magic School Bus Plants Seeds*. New York: Scholastic, Inc.

Cone, M. 1996. *Squishy, Misty, Damp & Muddy: The In-Between World Of Wetlands*. San Francisco: Sierra Club Books for Children.

Gibbons, G. 1991. *From Seed to Plant*. New York: Holiday House.

Johnson, R.L. 2001. *A Walk in the Deciduous Forest*. Minneapolis: Lerner Publications Co.

———. 2001. *A Walk in the Desert*. Minneapolis: Lerner Publications Co.

———. 2001. *A Walk in the Prairie*. Minneapolis: Lerner Publications Co.

Jordan, H. J. 1992. *How a Seed Grows*. New York: HarperCollins Children's Books.

Kesselheim, A.S., B.E. Slattery, S. H. Higgins, and M.R. Schilling. 1995. *WOW! The Wonders of Wetlands*. St. Michaels, Maryland: Environmental Concern, Inc.

Luenn, N. 1995. *Mother Earth*. New York: Simon & Schuster Children's.

Maestro, B.C. 1993. *How Do Apples Grow?* New York: HarperCollins Children's Books.

Micucci, C. 1996. *The Life and Times of the Apple*. New York: Orchard Books.

Parnall, P. 1971. *The Mountain*. New York: Doubleday Company, Inc.

Radin, R.Y. 1989. *High in the Mountains*. New York: Simon & Schuster Children's.

Rockwell, A. 1999. *One Bean*. New York: Walker & Company.

Seuss, Dr. (T.S. Geisel). 1976. *The Lorax*. New York: Random House, Inc. (Also available on VHS, 1996. Directed by Hawley Pratt. Fox CBS.)

Silverstein, S. 1986. *The Giving Tree*. New York: HarperCollins Children's Books.

Turner, A. 1989. *Heron Street*. New York: HarperCollins Children's Books.

Williams, T.T. 1985. *Between Cattails*. New York: Macmillan Library Reference.

Yolen, J. 1995. *Letting Swift River Go*. New York: Little, Brown, & Company.

Section III

Aliki. 1976. *Corn is Maize*. New York: HarperCollins Children's Books.

Allen, J., and T. Humphries. 2000. *Are You a Ladybug?* New York: Kingfisher Books.

Cole, H. 1998. *I Took a Walk*. New York: Greenwillow Books.

Cole, J. 1996. *The Magic School Bus Gets Eaten*. New York: Scholastic, Inc.

Collard, S.B. 1998. *Our Wet World: Exploring Earth's Aquatic Ecosystems*. Watertown, Massachusetts: Charlesbridge.

Dennard, D. 2000. *Coyote At Piñon Place*. Washington, DC: Smithsonian Institute Press.

Gackenbush, D. 1981. *Little Bug*. New York: Houghton Mifflin/Clarion Books.

George, L.B. 1999. *Around the World: Who's Been Here?* New York: Greenwillow Books.

Glaser, L. 1994. *Wonderful Worms*. Brookfield, Connecticut: Millbrook Press.

Hornblow, L., and A. Hornblow. 1990. *Insects Do the Strangest Things*. New York: Random House, Inc.

Johnson, J. 1997. *Simon and Schuster Children's Guide to Insects and Spiders*. New York: Simon and Schuster Children's.

National Science Teachers Association

Lauber, P. 1994. *Who Eats What?: Food Chains And Food Webs.* New York: HarperCollins Children's Books.

Lobel, A. 1986. *Grasshopper on the Road.* New York: HarperCollins Children's Books.

Orr, R. 2000. *The Burrow Book: Tunnel Into A World Of Wildlife.* New York: Dorling Kindersley.

Raschka, C. 2000. *Thingy Things: Wormy Worm.* New York: Hyperion Books for Children.

Schwartz, D.M. 1997. *Underfoot.* Huntington Beach, California: Creative Teaching Press, Inc.

Silver, D.M. 1995. *One Small Square: Woods.* New York: W.H. Freeman Company.

 Aardema, V. 1981. *Bringing the Rain to Kapiti Plain.* New York: Dial Press.

Booth, D. 1997. *The Dust Bowl.* Buffalo, New York: General Distribution Services, Inc.

Cooney, B. 1985. *Miss Rumphius.* New York: Penguin Putnam Books for Young Readers.

Elkington, J., J. Hailes, D. Hill, and J. Makeower. 1990. *Going Green: A Kid's Handbook to Saving the Planet.* New York: Viking Children's Books.

Hall, Z. 1998. *The Surprise Garden.* New York: Scholastic, Inc.

Lehn, B. 1999. *What Is a Scientist?* Brookfield, Connecticut: Millbrook Press.

Miles, B. 1991. *Save the Earth: An Action Handbook for Kids.* New York: Alfred A. Knopf.

Rockwell, A., and H. Rockwell. 1982. *How My Garden Grew.* New York: Macmillan Publishing Co.

Rutten, J. 1999. *Erosion.* Chanhassen, Minnesota: This Child's World, Inc.

Ryder, J. 1976. *Simon Underground.* New York: HarperCollins Children's Books.

———. 1990. *Under Your Feet.* New York: Simon & Schuster Children's.

Silver, D.M. 1993. *One Small Square: Backyard.* New York: W. H. Freeman Company.

Steele, M.Q. 1989. *Anna's Garden Songs.* New York: Greenwillow Books.

Stille, D.R. 1990. *Soil Erosion and Pollution.* Danbury, Connecticut: Children's Press.

Titherington, J. 1990. *Pumpkin, Pumpkin.* New York: Morrow, William, & Co.

Van Allsburg, C. 1990. *Just a Dream.* New York: Houghton Mifflin Company.

Teachers' Resources

American Association for the Advancement of Science. 1989. *Science for All Americans.* New York: Oxford University Press.

———. 1993. *Benchmarks for Science Literacy.* New York: Oxford University Press.

———. 2001. *Atlas of Science Literacy.* Arlington, Virginia: National Science Teachers Association.

Barrow, L.H. 2000. *Science Fair Projects: Investigating Earthworms.* Springfield, New Jersey: Enslow Publishers, Inc.

Doran, R., F. Chan, and P. Tamir. 1998. *Science Educator's Guide to Assessment.* Arlington, Virginia: National Science Teachers Association.

Gibb, L. 2000. Second grade soil scientists. *Science and Children* 37(3):24-28.

Global Learning and Observations to Benefit the Environment. 1997. Soil investigation. In *GLOBE Program Teacher's Guide.* Washington, DC: Global Observations and Learning to Benefit the Environment. Available at *www.globe.gov/sda-bin/wt/ghp/ tg+L(en)+UP(soil/Contents).*

Graveel, J.D., and S. Fulk-Bringman. *Demonstrations in Soil Science.* West Lafayette, Indiana: Purdue University. Available at *www.agry.purdue.edu/ courses/agry255/brochure/ brochure.PDF* or at *www.agry.purdue. edu/dept-info.htm.*

Kalman, B., and J. Schaub. 1995. *Squirmy Wormy Composters.* New York: Crabtree Publishing Company.

National Research Council. 1996. *National Science Education Standards.* Washington, DC: National Academy Press.

National Science Teachers Association. 2000. *NSTA Pathways to the Science Standards, Elementary Edition.* Arlington, Virginia: National Science Teachers Association.

Osborne Conservation Guides. 1992. *Protecting Trees and Forests.* Tulsa, Oklahoma: Educational Development Corporation.

Russell, H.R. 1998. *Ten Minute Field Trips.* Arlington, Virginia: National Science Teachers Association.

Soil and Water Conservation Society. 2000. *Soil Biology Primer.* Ankeny, Iowa: Soil and Water Conservation Society.

Trowbridge, L.W., and R.W. Bybee. 1995. *Becoming a Secondary School Science Teacher.* Upper Saddle River, New Jersey: Prentice Hall.

Utah Agriculture in the Classroom. *Dirt: Secrets in the Soil Video and Teacher's Guide.* Logan, Utah: Utah State University. Available at *www.ext.usu.edu/aitc/pages/resource/ dirt.htm.*

Agencies and Organizations

Natural Resources Conservation Service

For more information about soil and water conservation education, or about resource management, contact the Natural Resources Conservation Service (NRCS) or your local NRCS office. NRCS is an agency of the U.S. Department of Agriculture (USDA). There are 3,000 NRCS field offices nationwide, one in nearly every county. Your local office is listed in the telephone directory under "United States Government, Agriculture."

Natural Resources Conservation Service
Educational Relations/Conservation
  Communications Staff
U.S. Department of Agriculture
P.O. Box 2890
Washington, DC 20013
www.nrcs.usda.gov (main page)
www.statlab.iastate.edu/soils/nssc
  (National Soil Survey Center)

NRCS publishes county-level soil surveys for the United States. These soil surveys are available from your local NRCS office. You can also request these and other educational publications from NRCS by calling 1-888-LANDCARE or e-mailing landcare@swcs.org.

- *Backyard Conservation* (Program Aid 1621, 1998). Available at *www.nhq.nrcs.usda.gov/CCS/ Backyard.html*.

- *Buffers: Common Sense Conservation* (Program Aid 1615, 1997, U.S. GPO 1997-576-666). Available at *www.nhq.nrcs.usda.gov/CCS/ Buffers.html*.

- *The Colors of Soil* poster (1999, Poster 999901)

- *Conquest of the Land Through 7,000 Years* (Agriculture Information Bulletin 99, 1999)

- *Soil Erosion by Wind* (Agriculture Information Bulletin 555, 1994)

- *What is a Watershed?* (Program Aid 420, 1999)

- *Working with Wetlands* (Agriculture Information Bulletin 672, 1994)

- *The Water Cycle* poster (Program Aid 1588, 1999)

- *Your Hometown Clean Water Tour* poster (Program Aid 1587, 1998)

U.S. Department of Agriculture

Other sources of soil information from USDA include:

Ag in the Classroom
U.S. Department of Agriculture
1400 Independence Avenue, SW
Stop 2251
Washington, DC 20005
202-720-7925
www.agintheclassroom.org

Agriculture Research Service
Office of Information
56701 Sunnyside Avenue
Beltsville, MD 20705
301-504-9403
www.ars.usda.gov

National Agricultural Library
10301 Baltimore Avenue
Beltsville, MD 20705
301-504-5755
www.nal.usda.gov

USDA Forest Service
Sidney R. Yates Federal Building
201 14th Street, SW
Washington, DC 20250
202-205-1760
www.fs.fed.us

USDA Office of Communication
Public and Media Outreach Center
1400 Independence Avenue, SW
Washington, DC 20250
www.usda.gov/agencies/ocpage.htm

Other Organizations

There are hundreds of private and nonprofit organizations involved in resource conservation and education. Many of these professional organizations publish educational materials on soil. To conduct a search for related organizations or references via Internet search engines, use keywords such as "soil science," "sciences of soil," "soil education," or "soil lesson plans."

Other helpful resources are county-level resource conservation or local government conservation organizations. Throughout the United States, local citizen-districts carry out conservation activities that help to improve water quality, reduce soil erosion, and improve wildlife habitats.

American Geological Institute
4220 King Street
Alexandria, VA 22302
703-379-2480
www.agiweb.org

Boy Scouts of America
1379 West Walnut Hill Lane
Irving, TX 75038
www.bsa.scouting.org

General Federation of Women's Clubs
1734 N Street, NW
Washington, DC 20036-2990
800-443-GFWC
www.gfwc.org

National Science Teachers Association

Girl Scouts of the U.S.A.
420 Fifth Avenue
New York, NY 10018
800-478-7248
www.gsusa.org

National Association of Conservation Districts
509 Capitol Court, NE
Washington, DC 20002
202-547-6223
www.nacdnet.org

National Gardening Association
1100 Dorset Street
South Burlington, VT 05403
802-863-5251
www.garden.org

National Science Teachers Association
1840 Wilson Boulevard
Arlington, VA 22202
703-243-7100
www.nsta.org

National Wildlife Federation
8925 Leesburg Pike
Vienna, VA 22184
703-790-4000
www.nwf.org

Project Food, Land & People
Presidio of San Francisco
P.O. Box 29474
San Francisco, CA 94129
415-561-4445
www.foodlandpeople.org

Project Learning Tree
1111 19th Street, NW, Suite 780
Washington, DC 20036
888-889-4466
www.plt.org

Project WILD
707 Conservation Lane, Suite 305
Gaithersburg, MD 20878
301-527-8900
www.projectwild.org

Soil and Water Conservation Society
7515 Northeast Ankeny Drive
Ankeny, IA 50021
515-289-2331
www.swcs.org

Soil Science Society of America
677 South Segoe Road
Madison, WI 53711
608-273-8095
www.soils.org